Make:
Family Projects
for **Smart Objects**

Make:
Family Projects
for Smart Objects

Tabletop Projects That Respond to Your World

John Keefe

MAKER MEDIA
SAN FRANCISCO, CA

Printed The United States of America
Published by Maker Media, Inc., 1160 Battery Street East, Suite 125, San Francisco, California 94111.

Maker Media books may be purchased for educational, business, or sales promotional use. Online editions are also available for most titles (safaribooksonline.com). For more information, contact our corporate/institutional sales department: 800-998-9938 or corporate@oreilly.com.

Publisher: Roger Stewart
Editor: Patrick Di Justo
Copy Editor: Linda Keefe
Proofreader: Elizabeth Welch, Happenstance
    Type-O-Rama
Designer and Compositor: Maureen Forys, Happenstance
    Type-O-Rama
Indexer: Valerie Perry, Happenstance Type-O-Rama
Cover photos by Lucas Saugen

September 2016: First Edition

Revision History for the First Edition

2016-0915 First Release

See oreilly.com/catalog/errata.csp?isbn=9781680451238 for release details.

Safari® Books Online

Safari Books Online is an on-demand digital library that delivers expert content in both book and video form from the world's leading authors in technology and business. Technology professionals, software developers, web designers, and business and creative professionals use Safari Books Online as their primary resource for research, problem solving, learning, and certification training. Safari Books Online offers a range of plans and pricing for enterprise, government, education, and individuals. Members have access to thousands of books, training videos, and prepublication manuscripts in one fully searchable database from publishers like O'Reilly Media, Prentice Hall Professional, Addison-Wesley Professional, Microsoft Press, Sams, Que, Peachpit Press, Focal Press, Cisco Press, John Wiley & Sons, Syngress, Morgan Kaufmann, IBM Redbooks, Packt, Adobe Press, FT Press, Apress, Manning, New Riders, McGraw-Hill, Jones & Bartlett, Course Technology, and hundreds more. For more information about Safari Books Online, please visit us online.

How to Contact Us

Please address comments and questions concerning this book to the publisher:

Maker Media, Inc.
1160 Battery Street East, Suite 125
San Francisco, CA 94111
877-306-6253 (in the United States or Canada)
707-639-1355 (international or local)

Maker Media unites, inspires, informs, and entertains a growing community of resourceful people who undertake amazing projects in their backyards, basements, and garages. Maker Media celebrates your right to tweak, hack, and bend any Technology to your will. The Maker Media audience continues to be a growing culture and community that believes in bettering ourselves, our environment, our educational system—our entire world. This is much more than an audience, it's a worldwide movement that Maker Media is leading. We call it the Maker Movement.

For more information about Maker Media, visit us online:

- ▸ Make: and Makezine.com: makezine.com
- ▸ Maker Faire: makerfaire.com
- ▸ Maker Shed: makershed.com

To comment or ask technical questions about this book, send email to bookquestions@oreilly.com.

To Kristin, Kaia
and Natalie

About the Author

John Keefe is a maker, journalist, and professional beginner based in New York City. This book emerged from a year he spent trying to make something every week, posting each one on his blog at http://johnkeefe.net.

As you use this book, you can find updated information, code, corrections and links to parts at http://keefe.cc/family-projects.

You can email the author at familyprojects@johnkeefe.net and follow him on Twitter at @jkeefe.

About the Illustrations

All circuit diagrams in this book were made using Fritzing from http://fritzing.org.

All photographs, except for the cover image, by John Keefe.

Contents

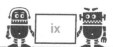

Welcome to the World of Smart Objects

The internet of things is a buzzphrase that's become almost meaningless. Is it a toaster that texts? Is it a fitness band on your wrist? Is it a wireless camera in an infant's room? Sure, it's all of those things and more.

These days, it's not much of a surprise that objects can sense things and communicate. You very likely have an uber-sophisticated sensor and communications system in your pocket or purse—capable of tracking your steps, capturing an image, and ordering your groceries. (I hear some people even use them to make phone calls!)

What might be surprising, is that it's not hard or expensive to make a sensing, communicating object yourself. And making smart objects can be rewarding, fun and even useful.

This book will give you the basics for building such things. The projects here are sometimes useful and sometimes just playful. With luck, they'll spark new ideas for how you might imagine, and build, your own useful and playful objects. Enjoy!

Acknowledgements

I couldn't have created "Family Projects for Smart Objects" without the love, support, and patience of Kristin, Kaia, and Natalie Keefe. Thank you from my heart. This was a family project made by you, too.

My thanks also to Team Blinky co-conspirators Liza Stark and Alex Goldmark for their boundless energy and maker-spirit, to Quinn Heraty for the encouragement and guidance I needed to become an author, to Eva Scazzero for trying all the projects and making them clearer, and to Patrick DiJusto for imagining that a book could be built from some of my blog posts—and then editing that book.

Also thanks to Linda Keefe, who raised me with a passion for projects, supported my curiosity even when I took apart household appliances and caught tons of typos in these pages.

—*John Keefe, July 2016, New York City*

Make:
Family Projects
for Smart Objects

You Only Have to Do This Once

You're about to embark on the world of smart objects using Arduinos. There's so much you can do and it's all a lot of fun. You'll say "Wow!" and "Yay!" and maybe even "Eureka!"

But first, we need to get you set up. This will take several steps, but I'll walk you through each one. Think of me as a computer professional (or a talented child) walking you through each step; a kind, patient computer professional/talented child.

Most importantly, you only have to do this once. I *really* want you to remember that, which is why I put it right up there as the title of this chapter. Once you're set up, the projects in this book will be a breeze.

So here we go!

Arduinos come in many, many flavors. That's largely because it's "open source" hardware, meaning anyone can build an Arduino from scratch. Each flavor has different features, sizes, and quirks.

Ingredients

- **1 Arduino Uno (Revision 3)**
- **1 Arduino USB cable**
- **Your computer: a desktop or laptop running Mac OS, Windows, or Linux**

I recommend you start out with the Arduino Uno, Revision 3. It's the classic beginner's board, and it's the board I'll assume you have throughout this book.

Almost any Arduino starter kit you find online will contain both the Arduino Uno and the Arduino USB cable. Many kits also have a bunch of the parts you'll need for the projects in this book. For a list of kits and ways to buy parts individually, and some key parts you won't find in a kit, check out Appendix A, "Everything You Need" at the end of the book, or visit http://keefe.cc/family-projects.

You also will need a computer onto which you have permission to install new software. If you've successfully installed new software before, you almost certainly have permission. If someone else such as a parent, a company manager, or the computer's owner has the Administrator password, you're going to need that person's help to continue.

NOTE ABOUT PARTS FOR LATER PROJECTS

In the last half of this book, we'll get your smart objects on the Internet—making them even smarter. Those projects require access to a wifi network and a particular wifi card for your Arduino. Get information about that part and other ingredients used throughout the book in Appendix A, "Everything You Need," or at http://keefe.cc/family-projects.

STEPS

To make the Arduino do things, we'll need to give it instructions. Normally, you give a computer instructions using a keyboard and a screen. But your Arduino doesn't have a keyboard or screen. So instead, we write the instructions on a desktop or laptop computer and send them to the Arduino over the USB cable.

So let's make that connection, physically linking your Arduino to your computer.

Get Connected

1. Plug the flatter end of the USB cable into a USB outlet on your computer.

2. Plug the squarish end of the USB cable into the matching outlet on the Arduino.

FIGURE 1-1: The squarish end of the USB cable goes into the squarish outlet on the Arduino.

3. If your computer alerts you about detecting a new device, just cancel out of the alert for now.

Great. You're connected, but you can't communicate with the Arduino until you install some free software onto your desktop or laptop computer. The installation is slightly different depending on your type of computer: Mac OS, Windows, or a type of Linux. So pick the next section that fits you, and remember: You only have to do this once.

For Macintosh Users (Mac OS X)

These are the steps for the Apple brand of laptops and desktop computers:

▶ **Get the free software.** Go to http://arduino.cc and click on the "Download" tab near the top of the page. There's a lot here, but just click the "Mac OS X" link near the top of the page to download the Arduino software for Mac.

NOTE FOR OLDER MACS

Playing with Arduino on an older Mac is a great use of a dusty computer. But you'll need to use an older version of the Arduino software. This means that my descriptions of the menu items, buttons, installation process, and other aspects of the software might not match your experience. Also, some of the more sophisticated projects later in the book might not work with the older software. So if your Mac system software is "Tiger," "Leopard," "Snow Leopard," or any version *earlier* than 10.7 ("Lion"), look on the Arduino download page for the "Previous Releases" section, click through, and download Arduino 1.0.6.

▶ **Optional contribution.** You'll be asked if you'd like to contribute money for the further development of Arduino software. That's up to you! (You can always come back to contribute later once you love Arduino.) Here, click either the "Contribute & Download" or "Just Download" link.

▶ **Find the downloaded file.** Your browser will download a file called `arduino-#.#.#-macosx.zip` (though instead of #'s you'll see numbers corresponding to the latest version of the software). This is a compressed, or "zipped," file containing the software. Your browser will probably put this file in your computer's "Downloads" folder. If you can't find the zipped file, use the "Spotlight" magnifying-glass icon in your computer's upper-right corner to search for "Arduino." Some browsers, like

Safari, may unzip the file for you. If you find a blue Arduino icon instead, you can skip the next step!

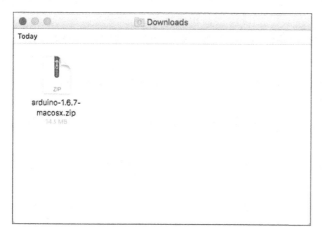

FIGURE 1-2: The Arduino .zip file sitting in the Mac OS X "Downloads" folder.

▶ **Unzip the downloaded file.** Once you find the `arduino-X.X.X-macosx.zip` file, double-click it. It will unpack the file and you'll get a blue Arduino application in the same folder. You no longer need the `arduino-X.X.X-macosx.zip` file and can throw it away if you want.

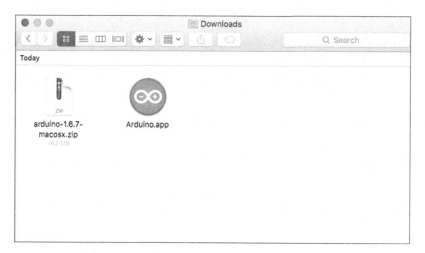

FIGURE 1-3: Welcome to the blue Arduino application icon!

▶ **Open the Arduino software.** Drag that blue icon to your computer's "Applications" folder. Then open your "Applications" folder and double-click the blue Arduino icon.

▶ **Pass through security.** You don't need to remove your shoes, but your Mac's security system may question you a bit. If you get an alert that you can't open Arduino "because it was not downloaded from the Mac App Store," click "OK" to clear the alert. To continue, you need to make a small change in your system settings. It's easy:

- Click the Apple in the upper-left corner of your computer's screen.

- Choose "System Preferences . . ."

- Pick the "Security & Privacy" icon (it looks like a little house).

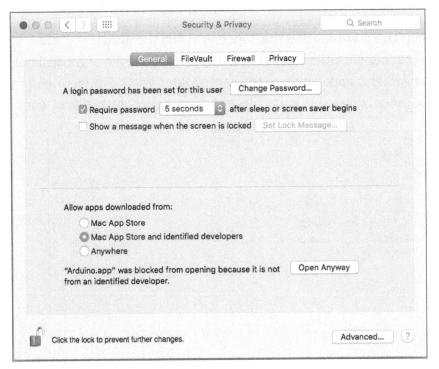

FIGURE 1-4: Adjusting the Mac's security settings to allow downloads from "Mac App Store and identified developers"

- If the lock at the bottom is closed, click it to allow changes here (you'll need to enter your computer's password).

- Under "Allow apps downloaded from:" choose "Mac App Store and identified developers."

- If you get an "Open Anyway" button, click it. Otherwise close this window and click on the blue Arduino icon again to launch the program.

▶ **Confirm that you're OK using your download.** You may get another alert warning that you're trying to launch something downloaded from the Internet. This one is easy to fix: just click "Open."

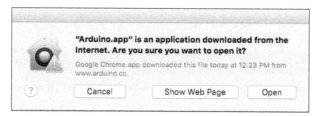

FIGURE 1-5: Sure you want to open it? Yes, please. (Just click Open.)

▶ **Answer one last question.** You might also get an alert asking whether you want to let Arduino accept incoming network connections. Here, click "Allow." Now skip ahead to the section "All Together Now."

For Windows Users

If your computer is running Microsoft Windows, this section is for you. The instructions might vary slightly based on the version of Windows you have.

▶ **Get the free software.** Go to http://arduino.cc and click on the "Download" tab near the top of the page. There's a lot here, but just click the "Windows Installer" link near the top of the page to download the Arduino software for your PC.

▶ **Optional contribution.** You'll be asked if you'd like to contribute money for the further development of Arduino software. That's up to you! (You can always come back to contribute later once you love Arduino.)

▶ **Download the software.** From the contribution page, click either "Contribute & Download" or "Just Download."

▶ **Find the downloaded file.** Your browser may start the installation once the file has been downloaded. If not, find the file in the "Downloads" folder. One way to get there is to click on *Start* and enter "downloads" in the search box. If the Arduino software isn't there, use the same search box to look for "Arduino."

▶ **Launch the downloaded file.** Once you find the Arduino file, double-click the blue Arduino icon to launch it.

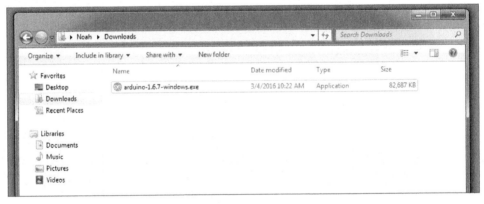

FIGURE 1-6: The Arduino software ready to install from the Downloads folder

▶ **Security check.** You may be asked if you're sure you want to install something downloaded from the Internet. Say yes!

▶ **License please.** Review the software license and click "Accept."

▶ **Options.** In the "Arduino Setup: Installation Options" window, leave everything checked and click "Next."

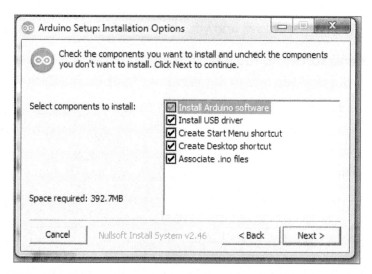

FIGURE 1-7: It's easiest to leave these checked as-is and click "Next."

▶ **Destination?** It's probably best to use the recommended destination folder and click "Install." The installation will begin, with a progress bar.

▶ **In Arduino we trust.** When you see "Would you like to install this device software?" first check the box "Always trust software from 'Arduino LLC.'" If you don't, you'll just be asked about this repeatedly. Careful now, because you need to click the "Install" button, *not* the "Don't Install" button, which is actually the one highlighted.

FIGURE 1-8: Check the box in this window and click "Install" (not the button that's highlighted.)

- **Almost there!** When you're done, "Arduino Setup Completed" will appear at the top of the install window, and you can click the "Close" button.

- **Start the Arduino software.** You'll now see the friendly blue Arduino icon both on your desktop and in the Windows Start menu. Click it!

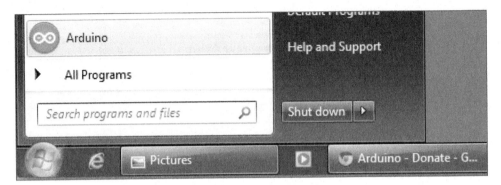

FIGURE 1-9: The Arduino software, loaded up and ready to go

- **Still sure?** The Windows security system may display one more alert as you get started. Click "Allow Access." You are ready to skip ahead to the section "All Together Now."

For Linux Users

Linux is an open source (and free) operating system. We have a computer at home running Ubuntu, one of the Linux flavors, and our kids use it all the time. That said, if you've ventured into Linux, I'm guessing you're pretty computer savvy. It takes some extra fiddling to run and manage Linux systems.

- **Get the free software.** Go to http://arduino.cc and click on the "Download" tab near the top of the page. There's a lot here, but just click the download link for the Linux type you use, either 32- or 64-bit. Definitely get the software from the Arduino website. We generally get our Ubuntu software from the built-in "Software Center"—but the version of Arduino I found there was *very* old. As I completed this book, Arduino was up to version 1.6.9. You'll want at least that.

▶ **Optional contribution.** You'll be asked if you'd like to contribute money for the further development of Arduino software. That's up to you! (You can always contribute later.)

▶ **Download the software.** From the contribution page, click either "Contribute & Download" or "Just Download."

▶ **Extract the software.** Find the downloaded file on your computer, probably in your "Downloads" folder, and extract that compressed file. In some systems, like our Ubuntu setup, you can view the contents of a compressed file without actually extracting it. Be sure to extract it.

▶ **Get to the command line.** This is where I'm going to assume that as a Linux user, you know a little more about your computer than most. At this point you need to get a command prompt. Usually, in Ubuntu and many other flavors, you do this by launching the "Terminal" program.

▶ **Navigate to the Arduino directory.** Using the command line navigate to the directory into which you extracted the downloaded file. If you extracted it directly in your "Downloads" folder, you'll type something like

```
cd ~/Downloads/arduino-#.#.#
```

▶ **Run the installer.** Once you're in the Arduino directory, run the installer shell script like this:

```
bash ./install.sh
```

That should do all the work of installing your software and putting a launching icon on your desktop.

ALL TOGETHER NOW

The software installation was quite possibly the hardest part of this book. You're almost done.

Arduino boards come in all sorts of shapes, sizes, and internal workings. We need to make sure the software knows which kind yours is and how it's connected to your computer. The software will make a good guess, but let's be sure it was right. Again, you only need to do this once.

▸ **Still connected?** Be sure your Arduino is still physically connected to your computer with the USB cable.

▸ **Set your "Board" type.** Since I'm assuming you're using an Arduino Uno, look to the Arduino software's menu bar and navigate to *Tools* ➜ *Board:*. You'll get a list of Arduino types. If it's not already checked, pick *Arduino Uno* or *Arduino/Genuino Uno* (Genuino is another name for Arduino; long story: https://en.wikipedia.org/wiki/Arduino#Legal_dispute).

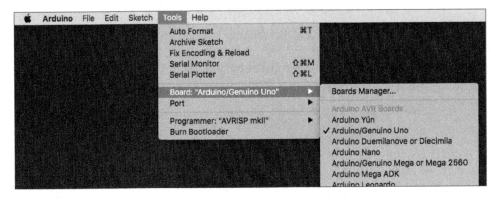

FIGURE 1-10: Here's where we tell the software which kind of Arduino you have. (The Windows and Linux versions are in the same place.)

▸ **Set your "Port."** We need to tell your computer where to find your Arduino. Navigate to *Tools* ➜ *Port:* and look at the drop-down menu. It'll look slightly different than mine, but you want the port that has "Arduino/Genuino Uno" in the name. On Macs, it probably starts with something like "/dev/cu.usbmodem . . . ." On Windows it'll start with "COM," and with Linux packages try the one that's "/dev/ttyACM0."

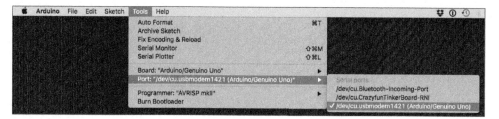

FIGURE 1-11: Telling my Mac where to find my Arduino. Yours may be slightly different. The Windows option probably starts with "COM," Linux with "/dev/ttyACM0."

If your Arduino Uno doesn't show up as an option, be sure your Arduino is connected to the computer with the USB cable. You may need to quit and relaunch the Arduino software. The software looks for Arduinos when it starts, and this should fix things.

If you ever have a moment when your Arduino just doesn't seem to be responding, check these settings. If they're not correct, the software will get cranky.

Finally, you may notice a "Programmer:" option in the *Tools* menu. With an Arduino Uno connected by a USB cable, as you have set up, the software ignores this setting—so you can, too.

Congratulations! You are ready to play with your Arduino. Give yourself a high-five and skip to the next chapter for your first project, "Hello Blinky World."

FIXES

If you have any trouble with the setup, rest assured you are not alone. Someone has almost certainly been in your exact same situation.

To get you on the way, search for your situation in Google or directly in the search box at http://arduino.cc—where you can also post your problem to the huge community of Arduino fans. I've also put helpful links to more installation instructions at http://keefe.cc/only-once.

Hello Blinky World

I f you've never made an Arduino do blinky things, you are so lucky! You get to experience that excitement for the first time.

Traditionally, when people begin learning something new about computers, their first project is to make the computer say, "hello world." Among blinky makers, our "hello world" is making an LED flash on and off. So that's what we'll do now.

Ingredients

- **1 Arduino Uno (Revision 3)**
- **1 Arduino USB cable**
- **1 LED**
- **Your computer**

CONCEPTS: USING YOUR ARDUINO

In this chapter, we'll write some instructions in Arduino code and then send them over to the Arduino.

You'll also learn a little bit about light-emitting diodes, or LEDs, which are fun little lights with a quirk you need to learn about.

This chapter, and all of the other chapters, assumes that you have already hooked up your Arduino Uno, downloaded the Arduino software for your computer, and gotten that software up and running. If you haven't already done that, jump back one chapter to "You Only Have to Do This Once" (which is true).

The Arduino, Arduino USB cable, and LED all come in almost any Arduino starter kit. You can also buy each part individually, if you wish. For links to kits and parts for this project, visit http://keefe.cc/hello-blinky.

STEPS

Wire Up the Parts

Besides the Arduino and its cable, there's just one other part to this project: the LED, or light-emitting diode.

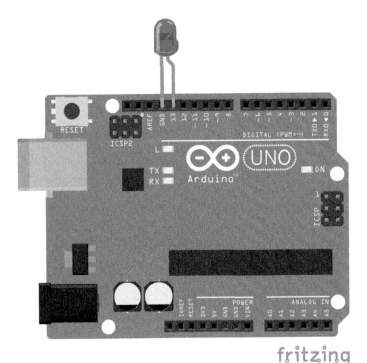

FIGURE 2-1: LED, meet Arduino: the wiring diagram for your first project

We'll use these little lights a lot, so you need to know about their one little quirk. Look carefully at the two "legs" of the LED and you'll see that one is longer than the other. That's because, like a battery, an LED has a positive "+" side and a negative "−" side. And, also like a battery, an LED doesn't work when it is installed backward.

Here's an easy way to keep the two sides straight: The "+" leg has had something *added* to it, so it is longer. The "−" leg has had something *subtracted* from it, so it is shorter. Keep that rule in mind, and you'll never be confused about how to attach an LED.

With that in mind, let's put the LED to use.

1. Along the edges of the Arduino there are two rows of holes, each of them labeled. Look for the one marked 13. Slide the LED's *longer* "+" leg into that hole.

2. Right next to hole 13 is one marked GND for "ground." Throughout this book, and in most electronic projects, "−" and "ground" are closely linked—and often the very same thing. In this case, we want to slide the LED's *shorter* "−" leg into the GND hole.

"HOLES" VS. "PINS"

The rows of holes on an Arduino are clearly rows of holes. But in electronics, these connection points are usually called "pins"—whether they are little pins or little holes. In books, online, and anywhere you read about Arduinos, writers call the holes *pins*. So from now on, I will, too . . . except in a couple of spots where it's just too confusing to call them "pins"!

At this point, your LED may already start blinking on and off slowly—changing every second or so. If that happens, it's because new Arduinos often come with the "hello world" blinking instructions already installed.

Whether or not your LED is blinking, let's go ahead and install a fresh copy of the "hello world" instructions, or program, to your Arduino.

Load Up the Code

It would be excellent if Arduinos understood instructions written in plain English (or any other human language). We could type:

```
Hi, Arduino!
There's an LED in hole 13.
Turn it on for 1 second.
Then turn it off for 1 second.
Keep doing that forever.
Thanks!
```

But, alas, we need to write the same thing in the Arduino language:

```
// the setup function runs once when you press reset or power the board
void setup() {
  // initialize digital pin 13 as an output.
  pinMode(13, OUTPUT);
}
// the loop function runs over and over again forever
void loop() {
  digitalWrite(13, HIGH);    // turn the LED on (HIGH is the voltage level)
  delay(1000);               // wait for a second
  digitalWrite(13, LOW);     // turn the LED off by making the voltage LOW
  delay(1000);               // wait for a second
}
```

Don't worry if that's complete gibberish; you don't need to speak Arduino to complete this or any other project in this book. If you *are* interested in coding, playing with smart objects is a great way to learn! For you, I've included a "Code Corner" section to every project, pointing out some interesting features of the code we're using.

And also don't worry about typing that code above, because it actually comes with the Arduino software you installed in the last chapter. Let's find it:

▶ If you haven't already done so, start your Arduino software by clicking on the blue Arduino application icon in your "Applications" or "Programs" folder.

▶ Navigate to the "Blink" sketch by going to the menu bar and choosing *File* ➜ *Examples* ➜ *01.Basics* ➜ *Blink.*

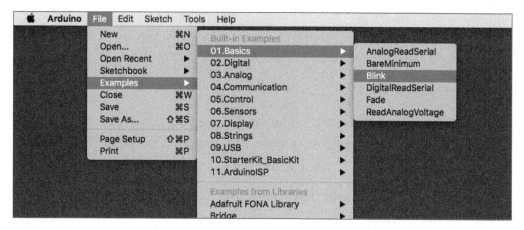

FIGURE 2-2: Finding the "Blink" example sketch on a Mac (it's in an identical spot on Windows and Linux versions).

Let's "upload" this code to the Arduino over the cable. You can do this from the menu by choosing *Sketch* ➜ *Upload,* but an even easier way is to click on the upload arrow at the top of the blue Arduino software window:

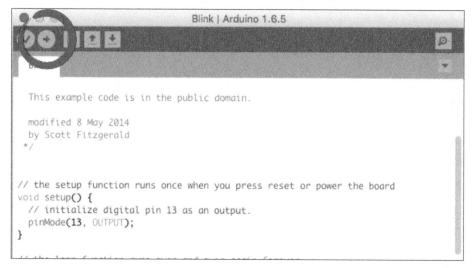

FIGURE 2-3: The magic "upload" arrow sends your program to the Arduino.

You should see the "Done uploading" message in the lower part of your Arduino window and your LED should be flashing (even if it was before).

Congratulations! You've just programmed your Arduino.

Fixes

Not working? Let's solve that.

There's usually a blue stripe right under where the code is written in the Arduino program window. If that turns orange and you get a message like "Problem uploading to board" . . .

- **Check your "Board" and "Port" settings.** By far the most common reason for this problem is that the board and port settings need to be adjusted under the *Tools* option in the menu bar. Jump back to the previous chapter for details pertaining to your type of computer. If the proper *Board:* or *Port:* options aren't available, you may need to quit your Arduino software and relaunch it so your Arduino board gets detected.

- **Check the USB cable.** Be sure the ends of your Arduino USB cable are pushed all the way into the sockets on both your Arduino and your computer.

If you uploaded successfully, and see a "Done uploading" message, but the LED isn't blinking . . .

- **Is the LED backward?** Remember: longer leg goes in pin 13, shorter leg in GND.

- **Are the LED legs actually in the holes?** The "pin" holes are tiny, and it's easy to miss one.

- **Are the LED legs touching each other?** Make sure they're not. That creates a "short circuit," and the power never reaches the glowing part.

- **Is the LED a dud?** It happens on rare occasions. Try another.

WHAT'S GOING ON?

Even without understanding Arduino language, you can tease out some of what's happening in the code—especially if I tell you that everything in the loop() section, set off by curly braces, just keeps repeating.

```
void loop() {
  // Everything here repeats forever.
}
```

This instruction turns the LED *on*:

```
digitalWrite(13, HIGH);
```

This turns the LED *off*:

```
digitalWrite(13, LOW);
```

Think of HIGH as "high power" and LOW as "low power, so low that it's off." The other key instructions are the delay() lines:

```
delay(1000);
```

They pause the program.

So taken all together, the instructions say turn the LED on, then pause, then turn the LED off, then pause. And since everything in the loop() section keeps repeating, we jump to the start of the loop after we're through the second pause.

TAKING IT FURTHER

The delay() command measures the delay in milliseconds, which is 1/1000th of a second. So the value of 1000 in the delay() instruction means delay 1 second. A delay of 2000 would be 2 seconds long, and a delay of 500 would be half a second. What happens if you change both of the delays to 500? Try it!

After you make your change, send the new program to the Arduino using the "upload" button or *Sketch* ➜ *Upload*. The change in the LED's behavior should be visible in just a few moments. What happened?

Play with the delay numbers. Make them small, large, and even different from each other. Just be sure to upload your program to the Arduino after each change. Can you predict what your change will do?

CODE CORNER

Interested in coding? Arduinos are a great way to learn! For some people, using code to control physical objects makes a lot more sense than keeping all of that code locked up inside a computer. "Code Corner" is the section in each project that helps you learn a little bit about the code we're using.

That said, *you are free to ignore the code*. Really. Focusing on the jumper wires and parts is a completely valid way to use this book. As long as you know how to get the code into your Arduino, you'll be set.

(But learning to code with physical objects is fun!)

Setup and Loop

This "Blink" program shows very nicely the basics of an Arduino program. First, you have the void setup() section wrapped in curly braces { and }:

```
void setup() {
  // initialize digital pin 13 as an output.
  pinMode(13, OUTPUT);
}
```

Everything in the void setup() section runs only once—when you power up the Arduino or when you press the reset button on the surface of the Arduino board.

In this case, the only thing that happens in the setup section is that we tell the Arduino to use Pin 13 as an output. That's because we want to send power *out* Pin 13 to the LED. The alternative is to set the pin as an input, which would allow us to read the electricity *in* to the Arduino from, say, a sensor.

As I mentioned above, everything in the `void loop()` section, between the curly braces { and }, runs over and over again.

Comments Are for Humans

Scan the "Blink" code and you'll see several lines with double slashes // in front of them, like this:

```
// turn the LED on (HIGH is the voltage level)
```

The double slashes mean "the words that follow on this line are for humans only." These are called "comments" and help people understand what's going on inside the program. The Arduino software ignores comments completely. In fact, comments aren't even sent to the Arduino (so they don't take up any of the Arduino's memory).

Another way to mark words as comments is to use /* at the beginning of the comment and */ at the end. This is useful for comments that take many lines, like this:

```
/*
 * These are comments for humans.
 * They will be ignored by the Arduino software.
 * Comments are good for understanding what's going on.
 */
```

I'll try to make sure all of the code in this book is commented well. That'll help you "read" the instructions we're giving to the Arduino, even if you don't speak Arduino.

YOUR NEXT PROJECT

The rest of the projects are ready for you now. They tend to build on each other through the book, so I recommend doing some of the next five projects first. After that, feel free to jump around!

If you get stuck, don't worry: There is an incredibly good chance that someone else has been stuck in exactly the same way. Google around using "Arduino" along with phrases and errors that describe your problem, or jump over to the Arduino forums at https://forum.arduino.cc/. Search for your problem, or if nothing matches, go ahead and ask the community for help.

A Dark-Detecting Light

Let's make our objects just a little smarter by adding the ability to notice changes in the immediate area with a sensor. In this case, we'll make our blinky bulb respond to the amount of light the sensor "sees."

Along the way, you'll learn how easy it is to detect light levels, and about the breadboard—a piece of plastic used to connect electronic parts together without soldering.

CONCEPT: SENSING LIGHT

Light sensors are among the simplest sensors around, and they're super cheap, too. The light sensor we'll use is a "photoresistor," a little piece of material that lets electricity pass through easily when light is present but resists electricity when there's little or no light.

Ingredients

- 1 Arduino Uno
- 1 Arduino USB cable
- 1 breadboard
- Your computer
- 1 LED
- 3 jumper wires
- 1 photoresistor (some kits call these a "photocell")
- 1 10K-ohm (10KΩ) resistor, which has a brown-black-orange stripe pattern

This is a very common method to sense things: Find materials and sensors that change their resistance to electricity when conditions change. Arduinos are really good at reading those changes in electrical flow and they can turn that information into numbers we can use.

Our project will use that information to detect darkness, flashing our LED fast and urgently when the light in the room is low—and, conversely, slow and calm when the room is bright.

All of these parts—including the photoresistor—come in most Arduino starter kits. For links to that kits, or parts you can buy individually, visit http://keefe.cc/dark-detector.

Before we jump in, let's make sure we understand those gazillion holes on the breadboard—because nothing quite makes sense unless we do.

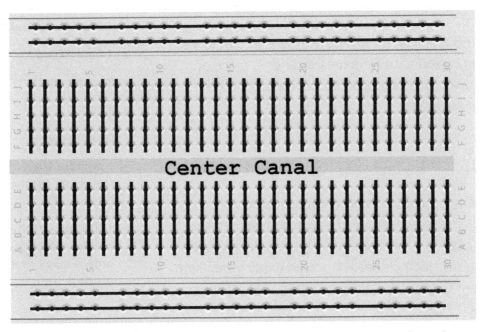

fritzing

FIGURE 3-1: The invisible connections in a breadboard, revealed!

The breadboard's holes are connected to one another in a pattern you can't see by just looking at it:

Notice that the holes in Row 1 from *a* to *e* are linked, as are the ones from *f* to *j*. But the two halves of Row 1 are *not* connected across the center canal, and they also aren't connected to any other rows. This is key to understanding how to use the breadboard.

Also the holes along the long, blue and red lines are connected to each other, but none of the "blue" holes are connected to any of the "red" holes. These lines are known as "power rails." They're useful for connecting power and ground down the length of the entire board, making it easy to reach from any of the rows. More on that later.

OK! Now we're ready.

STEPS

Wire Up the Parts

Here are the steps to put everything together:

- ▶ Grab the LED. As you learned last chapter, one leg is longer than the other. Put the longer LED leg into Pin 13 on the Arduino (remember, on an Arduino, a pin is actually a hole).

- ▶ Insert the shorter LED leg into the GND, or ground, hole right next door.

- ▶ Insert one of the photoresistor legs into Row 1 of the breadboard at Column *a*, and insert the other into Row 3 at Column *a*. Photoresistors don't have polarity, so it doesn't matter which leg goes in which row.

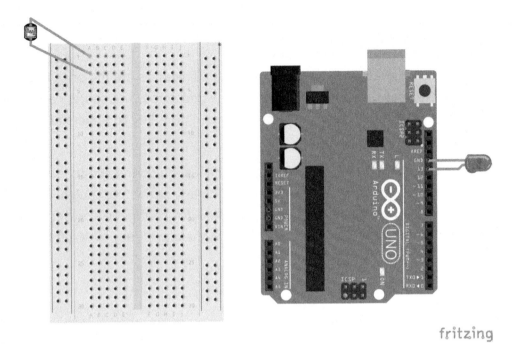

fritzing

FIGURE 3-2: The parts in place so far.

▸ Bring power to the photoresistor by connecting a jumper wire from the 5V pin on the Arduino to another hole in Row 1 on the breadboard. Remember, no electricity passes over the breadboard's center canal, so stay on one side of that canal. I used Column e. If you have a red jumper wire use that. Red wires are typically used for the connection to power, and it's a good habit to get into.

▸ Insert one end of a jumper wire into a hole in Row 3 on the breadboard. Again, we need to stay on the same side of the center canal, so I used breadboard Column e. Insert the other end of this wire into Pin A0 on the Arduino. In the illustration, I used a yellow wire. This connection will be used to measure the electricity flow.

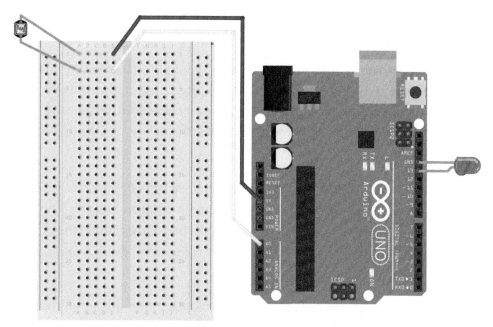

FIGURE 3-3: Project wiring up to this point

Let's pause a moment. If you're keeping close watch, it may seem like we're all set at this point: We're sending 5 volts of power out of the Arduino, through the photoresistor and then back into the A0 pin on the Arduino to measure the how much electricity has flowed through the circuit. So we should be able to detect changes in light, right?

Actually, not quite.

It turns out that we need to give the electricity a path to "ground," or the minus "−" side of the circuit. Not only that, but the path to ground needs a little something in the way so not *all* of the current goes to "ground" (which it will do if you let it). To make sure some power still goes to A0, we put a resistor in the route headed toward ground.

FUN FACT ABOUT GROUND

In your house, and in any device powered by your home's electricity, the "ground" part of the circuit is actually, eventually, connected to the real ground—as in outside under our feet! For battery-powered circuits, the negative "−" side of the battery serves as the ground.

The resistor we connect to ground is a static resistor—its resistance doesn't change. It looks like a small piece of plastic, with colored bands going all the way around. In some kits, they come in a little bag; other times they come taped together along the legs. If they're taped together, detach and use just one.

The bands, or stripes, on a static resistor are coded markings that indicate its resistance value. Right now, and in most of the other projects in this book, we want a 10K-ohm (10KΩ) resistor—whose code is brown, black, orange. There's often one more stripe at the end that is gold or silver. For a great resistor-color decoder, check out http://worrydream.com/ResistorDecoder/.

FIGURE 3-4: The brown-black-orange (and gold) stripes on a 10K-ohm (10KΩ) resistor

So, to finish up the wiring:

▶ Grab the static resistor, bend the legs at right angles, and put one leg into another hole on Row 3 on the breadboard. I used Column c. It doesn't matter which leg you pick; resistors don't have polarity.

▶ Bend the other leg of the resistor into another row, such as Row 7.

▸ Connect a jumper wire from Row 7 of the breadboard to any of the GND, or "ground," holes on the Arduino. Typically, and throughout this book, black jumper wires are used for the connection to ground.

▸ Double-check that all of your connections are on the same side of the breadboard canal!

When you're done, it'll look like this:

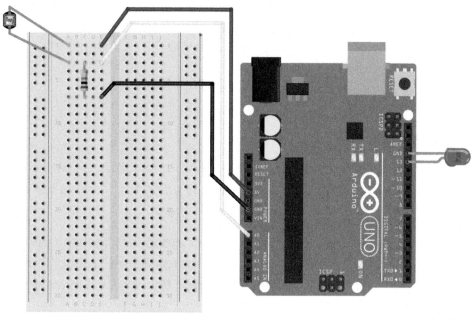

fritzing

FIGURE 3-5: Wiring diagram for the Dark-Detecting Light.

This setup—power to sensor, and then sensor to both a ground *and* a measurement hole—is incredibly common when you're working with Arduinos and sensors. So here's a little chart to come back to if you need a reminder later:

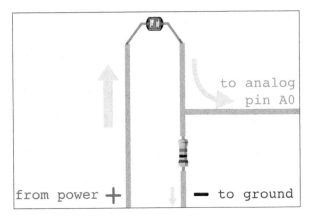

FIGURE 3-6: A common sensor circuit, with power going through the sensor to both a measuring point and to ground.

Load Up the Code

Once again, the code we'll use comes with your Arduino software, so it's easy to load up. Navigate to the "AnalogInput" sketch by going to the menu bar and choosing *File* ➜ *Examples* ➜ *03.Analog* ➜ *AnalogInput*.

When it's on your screen, upload the code to your Arduino by pressing the arrow button at the top of the blue window or using *Sketch* ➜ *Upload*. Check the "Load Your Code" section of the previous project if you need a refresher.

Make It Go

After a bunch of quick flashes of the LED, the program will start up and your LED will blink at a steady pace. Cover the photoresistor with your hand and see what happens.

The LED should blink at a fast, urgent pace when it's dark around the photoresistor, and should slow down when there is light present. Aim the photoresistor at a bright window or get it close to a bright bulb and it should slow down even more.

Fixes

If that's not happening, check your steps, being mindful these things:

▸ The LED's legs are different lengths. The longer one goes into hole 13. The shorter one goes into GND.

▸ No electricity flows across the breadboard canal. Be sure all of your connections are on one side.

▸ The wires and legs should be pushed firmly into the breadboard.

▸ It's easy to slip the legs into the wrong row on the breadboard. If you're following the wire-color conventions, you should see . . .

   ▪ A photoresistor leg and the red jumper wire in Row 1

   ▪ The other leg of the photoresistor, a leg of the static resistor, and a yellow jumper wire in Row 3

   ▪ The other leg of the static resistor and the black jumper wire in Row 7

WHAT'S GOING ON?

Even if you've never read code before, you can sort of see what's going on—especially because there are helpful human notes following the // characters.

The code inside the brackets labeled `void loop()` will repeat forever until you pull the power from the Arduino.

Just like the last chapter, the on and off pattern is paused by the `delay()` command. But *this* time, instead of a static number like 1,000, the delay changes depending on the value from the sensor, which is represented by `sensorValue`.

Let's walk through this. In low light . . .

▸ The photoresistor doesn't conduct electricity as well (it's more resistant), so *less* electrical voltage gets back to the Arduino's A0 pin.

- That gets detected as a *low* sensorValue number.

- Which means *less* of a delay between the LED's on and off states.

- Which makes the LED flash *faster*.

And in bright light . . .

- The photoresistor conducts electricity better (it's less resistant), so *more* electrical voltage passes through it and back to the Arduino.

- That registers as a *high* sensorValue number.

- Which means *more* of a delay where there's delay(sensorValue) in the code.

- Which makes the LED flash *slower*.

CODE CORNER

Take a look at the code we're using, and you'll see a little bit more than just the void setup() and void loop() sections. Here's the whole program:

```
/*
Created by David Cuartielles
modified 30 Aug 2011
By Tom Igoe
This example code is in the public domain.
http://www.arduino.cc/en/Tutorial/AnalogInput
*/
int sensorPin = A0;     // select the input pin for the potentiometer
int ledPin = 13;        // select the pin for the LED
int sensorValue = 0;    // variable to store the value  from the sensor
void setup() {
   // declare the ledPin as an OUTPUT:
   pinMode(ledPin, OUTPUT);
}
void loop() {
   // read the value from the sensor:
   sensorValue = analogRead(sensorPin);
   // turn the ledPin on
```

```
    digitalWrite(ledPin, HIGH);
    // stop the program for <sensorValue> milliseconds:
    delay(sensorValue);
    // turn the ledPin off:
    digitalWrite(ledPin, LOW);
    // stop the program for for <sensorValue> milliseconds:
    delay(sensorValue);
}
```

At the very top, we have a block of comments set off by /* and */. Again, those are ignored by the Arduino software.

And then right below the comment block, there's a section where several words get assigned numbers. Huh? Yeah. These are called *variables*. Here's the code in Arduino:

```
int sensorPin = A0;     // select the input pin for the potentiometer
int ledPin = 13;        // select the pin for the LED
int sensorValue = 0;    // variable to store the value  from the sensor
```

Ignoring the comments for a moment, here's the code in English:

```
Hey Arduino, when you see the word "sensorPin," use Arduino Pin A0 instead.
When you see the word "ledPin," use the number 13 instead.
And when you see the word "sensorValue," use the number 0 instead.
```

Two quirky things here. First, the `int` at the beginning of each line means the word will be storing an "integer"—a whole number like 1, 2, –3, or 0. The strange thing is that in Arduinoland pin numbers like `A0` are considered integers. Go figure.

The second thing is that these assignments aren't permanent; we can change them in the code. In the code for this project, `sensorPin` and `ledPin` don't actually change. But `sensorValue` does! It starts out as 0, but gets changed to the value read in from the sensor. This changeability is why these words are called *variables*.

TAKING IT FURTHER

Tinker with the Resistor

Try using static resistors of different values instead of the 10K-ohm (10KΩ) resistor. What happens? Can you explain why?

See the Sensor's Value

You can actually watch the value of `sensorValue` if you'd like. You just have to add two lines of code, which are easy to type in yourself. The first one is `Serial.begin(9600);` and it goes in the `void setup()` section, within the curly brackets, like this:

```
void setup() {
  // declare the ledPin as an OUTPUT:
  pinMode(ledPin, OUTPUT);
  Serial.begin(9600);  // < - - New line is here
}
```

The second one is `Serial.println(sensorValue);` and it goes within the curly brackets in the `void loop()` section, like this:

```
void loop() {
  // read the value from the sensor:
  sensorValue = analogRead(sensorPin);
  // turn the ledPin on
  digitalWrite(ledPin, HIGH);
  // stop the program for <sensorValue> milliseconds:
  delay(sensorValue);
  // turn the ledPin off:
  digitalWrite(ledPin, LOW);
  Serial.println(sensorValue); // < - - New line is here
  // stop the program for for <sensorValue> milliseconds:
  delay(sensorValue);
}
```

Type the lines *exactly* as they appear, being mindful of the capital letters (case matters) and the end-of-line semicolon.

► Upload the edited code to your Arduino.

► Open the Arduino software's Serial Monitor window from the menu bar: *Tools* ➜ *Serial Monitor*. This allows you to watch some of the communication from the Arduino itself.

► A window will appear with the `sensorValue` numbers! (If you see nothing, or see gibberish, be sure the menu at the bottom of the Serial Monitor window is set to "9600 baud.")

► Cover the photoresistor with your hand and see the numbers change.

Now you can see exactly how the Arduino is turning the light measurements into numbers!

Night Light

Night lights are awesome and useful, and the best ones save power by only lighting up when it's dark in the room. That's what we'll make with this project.

CONCEPTS: SENSING LIGHT, TAKING ACTION

We'll use a photoresistor to detect light. But this time, we'll also check the *amount* of light we're detecting and then *do* something with that information.

The circuit for this project is almost exactly the same as in "A Dark-Detecting Light." The code, however, is quite different. It includes "if-then" commands: *If* the room is dark, *then* send power to the light.

The concept of "if this happens, then take action" is very common in the code for smart objects, and programming in general.

Ingredients

- 1 Arduino
- 1 Arduino USB cable
- 1 Arduino power supply
- 1 breadboard
- 1 photoresistor (also called a "photocell")
- 1 LED (white, if you have one)
- 3 jumper wires
- Your computer

Optional Items

- 2 extension jumper wires (male-to-female)
- 1 hollow, translucent toy

Almost all of these parts come in most Arduino starter kits. The optional male-to-female jumper wires often do not, though. To get your hands on those, or any of the other parts in this project, visit http://keefe.cc/night-light.

As for the optional hollow toy, you can use anything that would glow with an LED inside. I'll use a plastic sheep, because that seems appropriate.

STEPS

Wire Up the Parts

Wire up the circuit just as I described in the previous chapter. If you've got a few different LEDs handy, feel free to pick a color you'd like for your night light. Come back here when you're done.

Great! You're ready to go.

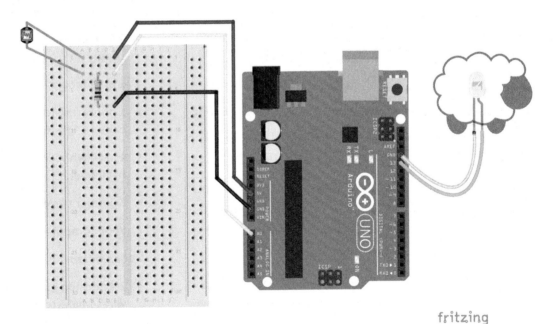

fritzing

FIGURE 4-1: Wiring diagram for the Night Light project, including optional extension cables and hollow toy

If you'd like to glow a toy instead of just your LED, follow these additional (and optional) steps:

- ▶ Take the LED out of its spot on the Arduino.

- ▶ Grab two of the "extension" jumper wires. These have an (actual) pin on one end, and a hole on the other. Often these wires come attached to each other in a ribbon, side by side. If yours came that way, peel off *two* jumpers at a time, so your two wires stay attached to each other.

- ▶ Put the pin ends of two jumper wires into the Arduino where the LED was (pins 13 and GND).

- ▶ Put the LED legs into the hole ends of the "extension" jumper wires, remembering that LEDs have polarity—the long leg needs to be wired back to Arduino Pin 13.

- ▶ Cut a hole in the hollow toy.

- ▶ Slide the LED inside.

- ▶ Use tape to hold it into place, if necessary.

If you have extra jumper wires, you can make a chain of them so the toy is even farther from the Arduino and breadboard. Just remember that the long leg of the LED needs to connect back to Arduino Pin 13.

Load Up the Code

Unlike the previous projects, the code for the Nght Light project doesn't come with your Arduino software (alas). But there are a few ways you can get it, and they're all free.

FROM THE WEB

I've set up a website full of useful links *and* the code you'll need for every project. Here's how to use that site to get what you need:

- ▶ Use a web browser to visit http://keefe.cc/night-light.

- ▶ Click the "Copy Code to Clipboard" button.

- ▶ Switch back to your Arduino software.

- ▶ Start a new Arduino sketch using *File* ➜ *New*.

- ▶ You'll get a mostly blank sketch window. Delete the little code that's there.

- ▶ Click anywhere inside the blank window.

- ▶ Paste the code you copied from the website using *Edit* ➜ *Paste* from the menu.

- ▶ Save your work using *File* ➜ *Save*.

FROM THE BUNDLE

Every sketch used in "Family Projects for Smart Objects" is also available as a zipped-up bundle you can download for free. Here's how to get the bundle.

- ▶ Use a web browser to visit http://keefe.cc/sketches.

- ▶ Your browser should start downloading the file right away.

- ▶ Once it's done, navigate to your "Downloads" folder.

- ▶ Unzip the file called `family-projects-sketches-master.zip`, which is usually done just by clicking or double-clicking its icon.

- ▶ Now you'll have a bunch of "chapter" folders with short names relating to a project in this book. Click on the `night_light` folder to open it.

- ▶ Inside, you'll see a file of the same name, but ending in .ino (for Arduino): `night_light.ino`. It'll have a blue Arduino icon.

- ▶ To use it, just double-click the file name and it should open in your Arduino software.

Now you're also set to run the code for every other project in this book!

FROM THE BACK OF THIS BOOK

The code for every project is also printed in the back as "Appendix B." If you're reading an electronic version of the book on your computer, you may be able to copy it from there and paste it into your Arduino software. This code happens to be pretty short, so you could actually type it, too.

- ▶ On your computer, start up your Arduino software and start a new sketch from the menu bar: *File* ➜ *New*.

- ▶ Delete what's in the window that pops up.

- ▶ Either type the code exactly as it appears in Appendix B or, if you have an electronic version of this book on your computer, copy-and-paste the entire code block for this chapter.

No matter how you get the code, once you have it save your sketch with a new name and then upload it to your Arduino by pressing the arrow button at the top of the blue window or using *Sketch* ➜ *Upload*. For more on how to get the code into your Arduino, including tips on what to do if the Arduino complains, see the "Load Your Code" section of the "Hello Blinky World" project.

At this point, the LED will stay off regardless of the light in the room. We still need to teach the Arduino what "dark" looks like where you are.

It's a Calibration

First, turn on the Serial Monitor on your Arduino software: *Tools* ➜ *Serial Monitor*. This lets you see communications from the Arduino.

You should see a new window with a bunch of numbers scrolling up. If you don't, be sure the drop-down menus at the bottom of the Serial Monitor window are set to "Both NL & CR" and "9600 baud."

The numbers scrolling by are the readings straight from the light sensor. The values should get smaller when you put your hand over the photoresistor, and larger when there's more light. In my setup, I see values around 780 in the bright room where I'm sitting, and values around 90 when I cup my hand over the sensor.

Your numbers will be different, but they should change noticeably when you shield the sensor from light. If they don't, walk through the wiring steps again, paying close attention to the holes you are using on the Arduino and on the breadboard.

Now we need to figure out what number is a good threshold for "dark" where you are. Make the room as dark as it'll be when the light should go on. (You also can also keep simulating a dark room by cupping your hands over the photoresistor; it's not hard to change the threshold later.)

With the sensor in the "dark," watch the numbers in the screen, and pick a number that roughly represents the point at which it's "dark." The LED will glow at any value *less than* the number you pick.

In the code, find this line near the top:

```
int darkPoint = 0;
```

Change the value after the equals sign from 0 to your "dark" value. Be sure to leave the semicolon.

Make It Go

You've changed the code, so upload it to the Arduino again (*Sketch* ➜ *Upload*).

Now, when your project is in the dark, the light should glow! When the lights are back on, it should go out again. If you need to adjust darkPoint, feel free to change the value and re-upload your code onto the Arduino.

You probably don't want your computer sitting next to your glowing toy all night, so once you're happy with the settings you can unplug your Arduino from the computer and plug the Arduino power supply into both the Arduino and a wall socket. Your program will restart once the Arduino has power again.

FIGURE 4-2: A sheep glows in the night!

Fixes

If the light doesn't reliably douse in the dark, it's possible that the glow from the LED is shining on the light sensor, tricking the photoresistor into thinking the room is still lit. Here are some possible fixes:

▶ Try to point the sensor away from the LED. Don't be afraid to bend the legs; just make sure they don't touch each other.

▶ If you're using the extension jumper wires, arrange your project so the LED is as far from the Arduino as possible.

- ▶ Make a tiny tube of dark tape or paper and place it around the sensor so it can only "see" light in one direction. Then point the tube and resistor away from the LED.

- ▶ Hook up your computer to the Arduino again and check the light values using the Serial Monitor (*Tools* ➜ *Serial Monitor*). See if there's a higher number that still represents "dark."

CODE CORNER

Remember, the code in the `void loop()` section just keeps running over and over again—checking the sensor value all day long.

In that loop, we use an "if-then" statement. There's a version of this statement in practically every computer language. In Arduino-speak, if-then statements look like this:

```
if (something here is true) {
    then do whatever is here
}
```

In the night light code, there are two spots where we do an if-then check.

If the room is dark, and the sensor reading is less than our "dark point," *then* send power to the LED, by setting Pin 13 to `HIGH`:

```
if (sensorValue < darkPoint) {
    digitalWrite(ledPin, HIGH);
}
```

If the room is bright, meaning the sensor value is greater than or equal to our "dark point," *then* don't power the LED, by setting hole 13 to `LOW`:

```
if (sensorValue >= darkPoint) {
    digitalWrite(ledPin, LOW);
}
```

You can put a lot of stuff between the brackets of an `if` statement, including more `if` statements! We'll see this used a lot in the projects to come.

TAKING IT FURTHER

Gently Fading Night Light

Whenever I make projects like this, I like to add some warm flair by actually fading up the LED instead of just turning it on abruptly (and gently fading it out, when daylight returns). This requires a little extra code to keep track of whether the LED is presently on or not and adjusting the if-then test slightly. In English, we might write this:

```
Take out a notebook to keep track of whether the LED is on or off.
Sense the light in the room.
If the room is dark and the LED is currently off then
    fade the LED to full power,
    and make a note that the LED is on.
If the room is bright and the LED is currently on then
    fade the LED to zero power,
    and make a note that the LED is off.
Repeat forever.
```

Notice that in the case that the room is dark and the LED is already on, we don't do anything. Because it's already on! Same for when the room is bright and the LED is already off.

To get the fading code for your project, visit http://keefe.cc/night-light.

Ice, Ice Blinky

Thermometers are important. They help you decide whether you need a jacket today, and they let you know if the aquarium water is the right temperature for your tropical fish. Thermometers connected to tiny computers can make decisions based on temperature, doing things like turning on lights at a certain temperature, sounding alarms when things are too hot or too cold, and keeping track of temperatures over time.

Let's get started by making an LED react to cold water.

Ingredients

- 1 Arduino Uno
- 1 Arduino USB cable
- 1 breadboard
- A piece of ice
- Your computer

Special Items

- 1 thermistor (model TMP36)

CONCEPTS: TEMPERATURE SENSING

"Thermistors"—you can think of the word as a mashup of "thermal" and "resistor"—are simple devices that change electrical resistance with the temperature. And as we've seen, Arduinos are excellent at detecting changes in resistance. So a thermistor and an Arduino make a very effective temperature sensor.

Almost all Arduino starter kits include a thermistor, and when they do it's usually the TMP36. This is a simple, black part with three legs and one flat side. If you don't have one, they're quite cheap. Find links to buying one at http://keefe.cc/ice-blinky.

STEPS

Wire Up the Parts

Here's how to wire up this project:

- ▶ Put the longer LED leg into Pin 13 on the Arduino.

- ▶ Put the shorter LED leg into the GND pin on the Arduino, right next door.

- ▶ Notice that the thermistor has three legs, and also that it has a rounded side and a flat side.

- ▶ Insert the three legs of the thermistor into in the breadboard's holes along Column *a* at rows 1, 2, and 3 *with the flat side facing the breadboard's printed row numbers*. The rounded side should be facing the other direction, toward the breadboard's center canal. You'll have to separate the legs a little to get them into each of the holes.

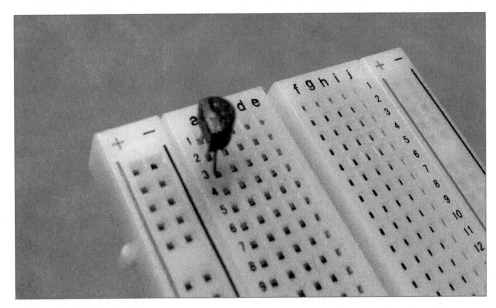

FIGURE 5-1: The position of the temperature sensor, with the flat side facing at the printed numbers along Column *a*

▶ Connect one end of a jumper wire (preferably a red one) to the Arduino's Pin 5V, and connect the other end to the breadboard's Row 1—let's use Column e.

▶ Connect one end of another jumper wire (any color; I used yellow) to the Arduino's Pin A0, and the other to the breadboard's Row 2 at Column e. Notice that we're staying on the same side of the breadboard's central canal.

▶ Connect one end of a (preferably black) jumper wire to any of Arduino's GND pins, and the other end of the wire to the breadboard's Row 3 at Column e.

▶ Double-check that all of your connections are on the same side of the breadboard's central canal.

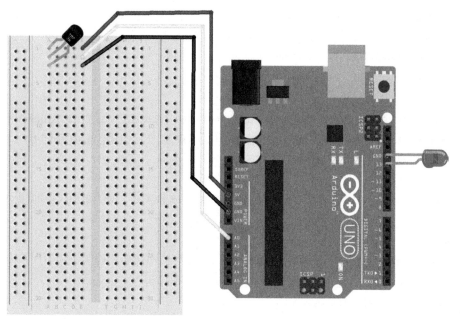

FIGURE 5-2: Wiring the temperature sensor to the Arduino. The flat side of the thermistor is pointing toward the left of this image, and the rounded side toward the breadboard's central canal.

Load Up the Code

Just as in "A Dark-Detecting Light," we'll use the example code that comes with your Arduino software. So with the Arduino software running, go to: *File ➜ Examples ➜ 03.Analog ➜ AnalogInput.*

Make It Go

Upload the code to your Arduino by pressing the arrow button at the top of the blue window or using *Sketch ➜ Upload.* The LED should start blinking. (If it doesn't, check the "Fixes" section below.)

Place a piece of ice on a dish next to the breadboard, carefully pick up the breadboard, and touch the thermistor to the ice. Watch the LED closely as you do. Can you see the flashing speed up?

Now pinch the thermistor between your fingers hold it tightly. Watch the blinking slow down as your body heat warms up the thermistor, causing the resistance to change in the other direction.

You've made a temperature sensor!

If you're having trouble seeing the flashing rate change, we can make a little adjustment to make it more pronounced. Look at the code and find this line:

```
sensorValue = analogRead(sensorPin);
```

Make one small change by adding *3 just to the left of the semicolon, like this:

```
sensorValue = analogRead(sensorPin)*3;
```

This will triple the sensor readings, making the warmer (higher) numbers bigger—and making the flashing change more noticeable.

Fixes

If your LED isn't lighting, be sure the correct leg is in the correct hole—remember, the longer leg goes into Pin 13.

If the LED is flashing but the speed doesn't change much, first be sure the wires are all connected correctly. Even if *nothing* is sensed, this circuit will flash the LED, but the rate of flashing won't change.

If the LED still isn't responding much, try swapping out the regular resistor for a value that's smaller. Try 10Ω instead. The base blink rate is tied to this resistor's value.

WHAT'S GOING ON?

Stick with me here, because how it works is kind of cool. (Ha.)

- ▶ The ice cools the thermistor.

- ▶ As the thermistor gets colder, it doesn't conduct electricity as well. (Another way to say that is that its resistance increases.)

- ▶ The Arduino senses how much electricity arrives back at the Arduino and gives it a number. That's the sensor value. (Technically, the sensor value is related to voltage.)

- ▶ With less electricity making it through the thermistor, the sensor value becomes a *smaller* number.

- ▶ That sensor value is used in the code as the number of milliseconds the program should wait between turning the LED on and turning it off (and then on again).

- ▶ When the sensor value gets *smaller* the wait between on and off gets *smaller*, too.

- ▶ If the delay between on and off (and on again) is smaller, the LED blinks *faster*.

TAKING IT FURTHER

Seeing the Sensor Reading

We can print the sensor reading into the Serial Monitor by adding two lines to our program. The first one is `Serial.begin(9600);` and it goes here:

```
void setup() {
  // declare the ledPin as an OUTPUT:
  pinMode(ledPin, OUTPUT);
  Serial.begin(9600);  // < - - Add this line
}
```

The second is `Serial.println(sensorValue);` and it goes within the curly brackets in the `void loop()` section, like this:

```
void loop() {
  // read the value from the sensor:
  sensorValue = analogRead(sensorPin);
  // turn the ledPin on
  digitalWrite(ledPin, HIGH);
  // stop the program for <sensorValue> milliseconds:
  delay(sensorValue);
  // turn the ledPin off:
  digitalWrite(ledPin, LOW);
  Serial.println(sensorValue); // < - - New line is here
  // stop the program for for <sensorValue> milliseconds:
  delay(sensorValue);
}
```

Type the lines *exactly* as they appear, being mindful of the capital letters (case matters) and the end-of-line semicolons. Or copy it from http://keefe.cc/ice-blinky.

Upload the code to your Arduino by pressing the arrow button at the top of the blue window or using *Sketch* ➡ *Upload*.

Then open the Serial Monitor using *Tools* ➡ *Serial Monitor* and watch the numbers change when you touch the sensor to the ice or pinch it between your fingers. You'll probably notice that the numbers here are *not* temperature readings. We can fix that.

But What's the Temperature?

If you're using a temperature sensor, a flashing light may be fun, but you probably want to know the actual temperature! I've provided a little code to do just that.

Start a new sketch using *File* ➜ *New*, delete what's there, and replace it with the little sketch below. It's short enough that you could actually type it in, but you could also copy-and-paste it from http://keefe.cc/ice-blinky.

```
#define SENSORPIN 0
float sensor_value;
float voltage;
float tempC;
float tempF;
void setup() {
  // Start the Serial Monitor stream
  Serial.begin(9600);
  pinMode(13, OUTPUT);
}
void loop() {
  sensor_value = analogRead(SENSORPIN);
  Serial.print("Sensor Value: ");
  Serial.println(sensor_value);

  voltage = sensor_value * 5000 / 1024;
  Serial.print("Voltage (millivolts): ");
  Serial.println(voltage);

  tempC = (voltage-500) / 10;
  Serial.print("Degrees Celsius: ");
  Serial.println(tempC);
  tempF = (tempC * 9/5) + 32;
  Serial.print("Degrees Fahrenheit: ");
  Serial.println(tempF);
  Serial.println();

  delay(5000);
}
```

Upload this new code to your Arduino by pressing the arrow button at the top of the blue window or using *Sketch* ➜ *Upload*.

Then open the Serial Monitor again using *Tools* ➔ *Serial Monitor* and watch the numbers change. This time, the sensor readings are actual temperature values!

A More Useful LED

Now that we know the temperature, we can make the LED actually useful by adding some "if-then" code. Let's say that *if* the temperature is above 80 Fahrenheit, *then* light the LED.

To do that, simply type the following lines of code immediately above the `delay(5000)` line:

```
if (tempF > 80) {
    digitalWrite(13, HIGH);
} else {
    digitalWrite(13, LOW);
}
```

If you still have your LED in the Arduino Pin 13, you're all set. It will glow whenever the temperature at the thermistor climbs over 80. Tinker with the "80" number to make it your own. If you prefer to use the Celsius value instead, replace `tempF` with `tempC`.

CODE CORNER

In and Out

A nifty feature of the Arduino is that all of the digital pins—those numbered 0 through 13 along one side—can send power *and* sense power. But an individual pin can't do both at the same time.

That's why at the beginning of this sketch, we declare the `ledPin` (which represents Pin 13) as an *output*. Like this:

```
pinMode(ledPin, OUTPUT);
```

In this case, we want to send power *out* to the LED. Later, when we use a digital pin to detect power coming *in* to the Arduino, we'll set the pin we want to use as an "input."

Float Along with Me

If you checked out the "Code Corner" from earlier chapters, you know a little bit about variables—words that can represent other values. You learned that you could make a variable represent an integer number, such as 1, 2, 13, and 0, by putting `int` in front of the variable name. Like so:

```
int sensorValue = 0;
```

Take a peek at the code in "But What's the Temperature?" above (even if you didn't actually use that code). Near the top, you'll see something curious. Instead of `int`, you'll see `float`.

```
float sensor_value;
float voltage;
float tempC;
float tempF;
```

The word `float` refers to numbers with decimal points like 1.5 or 200.333. These are "floating point" numbers, hence the `float` designation.

Variables with decimals require more space in the Arduino's tiny brain, and here the code is essentially saying "Hey Arduino, save some space for these four decimal numbers we'll call sensor_value, voltage, tempC, and tempF. We're not going to assign them now, but we'll be using them later."

Constant and Reliable

In that same code, you'll see another curious addition at the top:

```
#define SENSORPIN 0
```

The word "SENSORPIN" is being assigned to 0, much like we did with `int sensorValue = 0;`. But in this case, it's not allowed to change. It must always stay as 0. So it's not variable; it's a *constant*.

Why not just make it a variable? Because constants take even less of Arduino's brain power than variables do. Making "SENSORPIN" a constant is more efficient, which can become important when programs start getting complex enough to push the limits of the Arduino's little brain.

A Gentle Touch

When you tap a "button" on a smartphone's screen you obviously aren't pushing a button. The phone is detecting your finger on the glass. The same principles working on that screen to sense a finger can be used to make a touch-sensitive "button" with no moving parts!

CONCEPTS: DETECTING TOUCH, CAPACITANCE SENSING, ARDUINO LIBRARIES

The secret to the phone screen—and to this project—is the property known as *capacitance*. Capacitance is an object's ability to hold an electric charge, no matter how small.

Humans exhibit a fair amount of capacitance, as you can experience by shuffling your feet across a carpet and then touching a doorknob—the spark that jumps from your finger contains thousands of volts of electricity, which had been stored in your body!

Our devices can detect our body's little glow of electricity by noting a change in *capacitance* at the point where someone touches something.

Ingredients

- 1 Arduino Uno
- 1 Arduino USB cable
- Your computer
- 1 LED
- 3 jumper wires
- 1 resistor, with the highest resistance you have
- 1 square of aluminum foil, about the size of a cracker
- A piece of tape

There's a more complete description of capacitance if you search the Internet, but for the purposes of this project, just know that we will be sensing that electrical charge in the human body.

All of the items in this project, besides the aluminum foil and tape, should be available in any Arduino starter kit, and if you've completed any of the previous projects, you're all set for ingredients. For links to the parts, and to a variety of starter kits, visit http://keefe.cc/gentle-touch.

For the resistor, use the highest value you have. Note that with resistor values, *M* means million, *k* means thousand, and Ω means "ohms," the units of resistance. So 10MΩ is higher than 10kΩ, which is higher than 10Ω. Anything above a 50kΩ resistor should work, though higher is better.

The colored bands on static resistors indicate their resistance value. For a great decoder table, visit http://worrydream.com/ResistorDecoder/.

STEPS

Wire Up the Parts

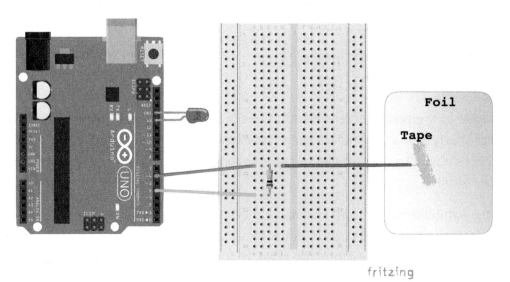

FIGURE 6-1: Wiring up the "Gentle Touch" project, with aluminum foil and tape.

Here it is in words:

- Insert the longer leg of the LED in the Arduino's Digital Pin 13.

- Insert the shorter leg of the LED into the next-door GND pin.

- Put one leg of the resistor into a numbered row on your breadboard.

- Into the same numbered row, insert one end of a jumper wire.

- Put the other end of that jumper wire into Digital Pin 6 of the Arduino.

- Into that same numbered row on the breadboard, insert one end of a second jumper wire.

- Tape the other end of that jumper wire to the aluminum foil, making sure the metal end of the wire is stuck to the foil.

- Back at the breadboard, put the second leg of the resistor into a *different* numbered row.

- Grab a third jumper wire and insert one end into this same row.

- Attach the other end of the third jumper wire to the Arduino's Digital Pin 4 (not the one marked A4, which is an analog pin).

You're ready to go.

Load Up the Code

The Arduino's brain is pretty tiny, and the list of commands it knows is small. So sometimes we need to add to its knowledge. For this, we turn to a library.

In programming, a "library" isn't as comprehensive as it sounds. It is a little bit of code containing commands specific to a particular task, such as sensing capacitance.

The Arduino software comes with a lot of libraries, but often you have to add one from the Internet. That's what we need to do now. So it's a good time to learn how to add libraries, and it's not hard. We'll be adding a library provided by Paul Stoffregen at https://github.com/PaulStoffregen/CapacitiveSensor.

▸ Download the "CapacitiveSensor" library by putting this short link to Paul's library into a browser like Safari, Chrome or Firefox: http://j.mp/touch-sense.

▸ That link should make your browser download a compressed file called CapacitiveSensor-master.zip, which is likely in your "Downloads" folder. We'll use this file in its compressed ".zip" form—so don't open it manually. Doing so will decompress it!

WAS YOUR FILE POLITELY DELETED?

Some Internet browsers, notably Safari for Macs, will "helpfully" open ".zip" files for you, decompress them, and then delete the original ".zip" file. But that's not helpful in this case! We need that original ".zip" file.

If you end up with an uncompressed "CapacitiveSensor-master" folder, and no ".zip" file, check your browser's preferences. See if you can politely ask it to *not* open (and delete) the ".zip" files you download. In Safari, for example, go to *Preferences* . . . On the "General" panel, uncheck the box next to "Open 'safe' files after downloading."

Another option is to re-zip that folder. On a Mac, find the file, click on it once to highlight it, and then use *File* ➜ *Compress "CapacitiveSensor-master"*. In Windows, it'll be similar, though you right-click on the folder first to highlight it. With Windows 10, for example, right-click on the folder and pick *Send to* ➜ *Compressed (zipped) folder*.

▸ Start your Arduino software (if it's not running already).

▸ From the menu bar, select *Sketch* ➜ *Include Library* ➜ *Add .ZIP Library* . . .

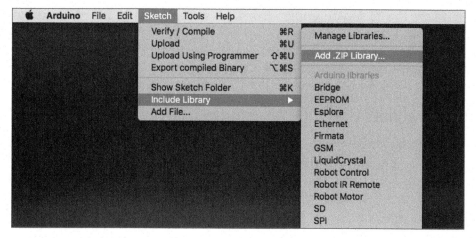

FIGURE 6-2: Adding a downloaded library on a Mac (it's in the same spot in the Windows version)

▶ Navigate to the place your browser saved `CapacitiveSensor-master.zip` (it's very likely in your "Downloads" folder).

▶ Select `CapacitiveSensor-master.zip` and click the "Choose" button.

▶ Great! Now quit your Arduino software and restart it.

Now that the library is in your Arduino's collection, we're ready to load in the project code. As outlined earlier, there are three ways to do that:

FROM THE WEB

To get the "Gentle Touch" code from this book's companion website:

▶ Use a web browser to visit http://keefe.cc/gentle-touch.

▶ Click the "Copy Code to Clipboard" button.

▶ Switch back to your Arduino software.

▶ Start a new Arduino sketch using *File* ➜ *New*.

▶ You'll get a mostly blank sketch window. Delete the little code that's there.

▶ Click anywhere inside the blank window.

▸ Paste the code you copied from the website using *Edit* ➔ *Paste* from the menu.

▸ Save your work using *File* ➔ *Save*.

FROM THE BUNDLE

If you don't already have the bundle, the steps for getting it are outlined at the beginning of Appendix B. (It's free!) Once you have the bundle . . .

▸ Find your `family-projects-sketches-master` folder, and double-click it to open it.

▸ Click on the `gentle_touch` folder.

▸ Inside, you'll see `gentle_touch.ino`. It'll have a blue Arduino icon.

▸ To use it, just double-click the file name and it should open in your Arduino software.

FROM THE BACK OF THIS BOOK

The code for this project is printed in the back of the book in Appendix B— and it's short enough to type. Hand-typing code is a good way to learn about what's going on. So feel free to enter it by hand, being certain to type it exactly as written and paying attention special attention the semicolons at the ends of some lines. Also watch the use of capital letters. Case matters, and this program uses both `CapacitiveSensor` and `capacitiveSensor`, which represent two different things!

If you're reading an electronic version of the book on your computer, you may be able to simply copy it from Appendix B and paste it into your Arduino software.

So the steps here are:

▸ On your computer, start up your Arduino software and start a new sketch from the menu bar: *File* ➔ *New*.

▶ Delete what's in the window that pops up.

▶ Either type the code exactly as it appears in Appendix B or, if you have an electronic version of this book on your computer, copy-and-paste the entire code block for this chapter.

No matter how you got the project code into your Arduino software, save your work, using *File* ➜ *Save*.

Make It Go

Upload the code to your Arduino by pressing the arrow button at the top of the blue window or using *Sketch* ➜ *Upload*. Then open the Serial Monitor using *File* ➜ *Tools* ➜ *Serial Monitor*. You should see numbers scrolling by.

Now touch the foil. What happens? Do the numbers jump? They should! The bigger the resistor you used, the higher the numbers will jump.

Can you detect a change in values by getting *really close* to the foil without touching it?

The LED should light up when you touch the foil, too. If it doesn't, we have a fix for that.

Fixes

If your LED isn't lighting when you touch the foil, your sensor might not be sensitive enough. (Or if it's stuck on, it could be too sensitive.)

The "if-then" action part of the code currently reads, in English, "If the sensor reading is greater than 100, send high power to the LED. Otherwise (else), send low power."

Here it is in Arduino:

```
if (sensor_reading > 100) {
 digitalWrite(LED, HIGH);
} else {
 digitalWrite(LED, LOW);
}
```

The sensor readings are the same values you saw scrolling past in the Serial Monitor, so pick a value that represents a "touch" for your sensor. Type that number in the code instead of the 100.

Once you change the value, don't forget to re-upload the code to your Arduino (Sketch ➜ Upload or use the arrow button at the top of the blue coding window).

If you're getting "0" readings in the Serial Monitor, you may need a stronger resistor. Use the highest value you have. The stripes on the resistor indicate its value. A great resource for decoding them is here: http://worrydream.com/ ResistorDecoder/.

WHAT'S GOING ON?

The instructions tell the Arduino to keep checking the capacitance levels on Pin 6, which is where we've attached our foil. The "if-then" section turns those numbers into actions.

CODE CORNER

It's super common in coding to use additional libraries to add features you don't have and that someone has already figured out. Once you added the CapacitiveSensor library to your Arduino software collection, it gets used in the project code with a simple line at the top of the program:

```
#include <CapacitiveSensor.h>
```

Then, on the next printed line, we put those extra smarts to use:

```
CapacitiveSensor   cs_4_6 = CapacitiveSensor(4,6);
```

This assigns the variable cs_4_6 to a new CapacitiveSensor skill, or function, from the library. To use that function, we insert 4,6 between the parentheses, which lets the library know what Arduino pins we're using to read the sensor.

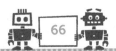

The value of the sensor then gets put into the `cs_4_6` variable as a number. Cool, right?

TAKING IT FURTHER

Instead of foil, you could use conductive fabric like copper taffeta. It's super fun to play with—and that it even exists just makes me smile. Some people who make wearable electronics use conductive fabric together with capacitive sensing to sew functioning "buttons" right onto their clothing!

For links to conductive fabric sellers, and examples that wire conductive patches to sewable Arduinos, visit http://keefe.cc/gentle-touch.

Someone Moved My Stuff!

Imagine an adventure movie where our hero is in search of the Object That Will Save Earth. After outwitting evil-doers, finding a hidden cave, and sneaking past laser-beam alarms, she puts her hands on the valuable object and picks it up. Moments later, red lights flash and alarms sound.

It is that moment we're going to re-create!

CONCEPTS: SENSING FORCE

Just as there are materials that change their conductivity with different light levels or temperature, there are also materials that change conductivity when something is pushing against them—like a finger, a toy or the weight of an Earth-saving object.

Ingredients

- 1 Arduino
- 1 Arduino USB cable
- 1 breadboard
- 5 jumper wires
- 1 pressure sensor
- 1 little buzzer
- 1 10k-ohm (10kΩ) resistor, which has a brown-black-orange stripe pattern
- a small object you cherish :-)
- Your computer

Optional part

- 1 Arduino power supply

We call these "force sensors," and they look like this:

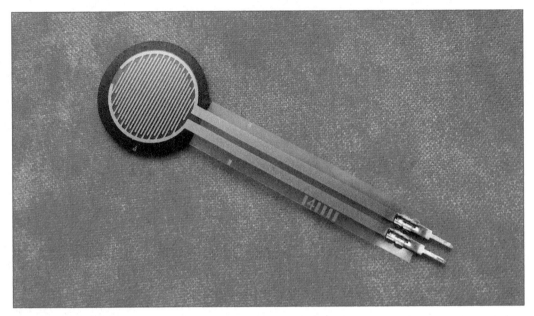

FIGURE 7-1: A simple sensor to measure pressure, or force

This will be our sensor for this project.

All of the parts you need come in almost every Arduino starter kit. If you don't have either the force sensor or the buzzer, there are links to places you can buy them at http://keefe.cc/stuff-alarm.

STEPS

Wire Up the Parts

As you make more projects using electronic parts, you'll begin to notice that many of the parts need to connect to "ground," which is GND on the Arduino. To make it easier to wire up all of these parts, the breadboard has long blue and red rows—which I call "rails"—running the length of the board. It's pretty

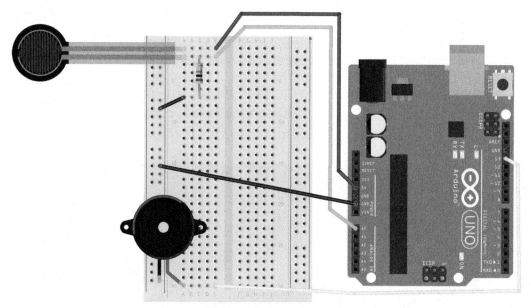

fritzing

FIGURE 7-2: Wiring chart for the force sensor project

common to connect a jumper wire from one of the blue rails, marked with a "–" minus sign, to the GND on the Arduino. That way, all of the parts needing ground can connect with a short hop to the rail.

Let's first connect our ground rail, and we'll go from there.

▶ Hold the breadboard so all of the numbers and letters are upright, and spot the blue row of holes on the left side of the board.

▶ Connect a jumper wire, preferably black for ground, from any GND pin on the Arduino to one of the holes along the blue row. (Why breadboard ground rails are marked blue instead of black is a mystery to me!)

For each of the following steps, use the holes on the same side of the breadboard's center canal as our ground rail. Remember that holes in a row connect to each other, except across that canal.

- ► Insert the "feet" of the force sensor into the top two rows of the bread-board—so one foot is in the top row and the other foot is in the second row. Be sure the feet are pushed as far in as they will go.

- ► With your finger right where the force sensor and the breadboard meet, gently bend the sensor contacts so they form a right angle. This way your sensor is closer to horizontal than standing straight up.

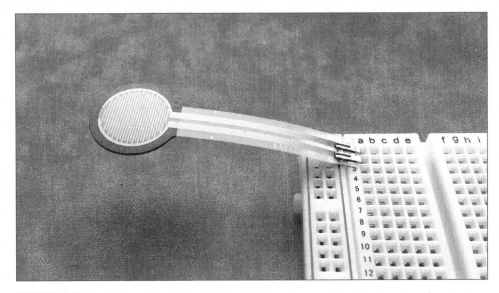

FIGURE 7-3: Gently bend the pins of the pressure sensor so the whole sensor is more horizontal than vertical.

- ► Connect a jumper wire (preferably red, for power), from the top row of the breadboard to Pin 5V on the Arduino.

- ► Connect another jumper wire (any color) from the second row of the breadboard to Pin A0 on the Arduino.

- ► Insert one leg of the resistor into the second row on the breadboard

- ► Insert the other leg of the resistor into a new row—let's use Row 7.

- ► Connect a jumper wire (preferably black, for ground) from Row 7 to the blue ground rail we wired up in the first step.

Grab the little buzzer from your kit. It's usually a black, round plastic part with a hole in it (where the sound comes out). Buzzers have polarity, like LEDs or batteries, so look for signs of that. The positive side may have a '+' plus sign, a red wire, or a longer leg. The negative side may have a black wire, a "−" minus sign, no sign, or a shorter leg.

- ▶ Insert the negative "−" side of the buzzer into a hole along the blue ground.

- ▶ Insert the positive '+' side of the buzzer into the bottom row of the breadboard (very likely, it's Row 30).

- ▶ Connect one more jumper wire (any color) from the same bottom row of the breadboard to Arduino Pin 13.

Load Up the Code

You've probably picked your favorite way to get code for these projects, but here are the three methods for this chapter's code:

FROM THE WEB

- ▶ Use a web browser to visit http://keefe.cc/stuff-alarm.

- ▶ Click the "Copy Code to Clipboard" button.

- ▶ Switch back to your Arduino software.

- ▶ Start a new Arduino sketch using *File* ➜ *New*.

- ▶ You'll get a mostly blank sketch window. Delete the little code that's there.

- ▶ Click anywhere inside the blank window.

- ▶ Paste the code you copied from the website using *Edit* ➜ *Paste* from the menu.

- ▶ Save your work using *File* ➜ *Save*.

FROM THE BUNDLE

If you don't already have the bundle, the instructions are at the beginning of Appendix B.

- ▶ Find your `family-projects-sketches-master` folder, and double-click it to open it.

- ▶ Click on the `stuff_alarm` folder.

- ▶ Inside, you'll see `stuff_alarm.ino`. It'll have a blue Arduino icon.

- ▶ To use it, just double-click the file name and it should open in your Arduino software.

FROM THE BACK OF THIS BOOK

The code for this project isn't too long, and it's printed in the back of the book. If you're reading an electronic version of the book on your computer, you may be able to simply copy and paste it into your Arduino software:

- ▶ On your computer, start up your Arduino software and start a new sketch from the menu bar: *File* ➜ *New*.

- ▶ Delete what's in the window that pops up.

- ▶ Either type the code exactly as it appears in Appendix B or, if you have an electronic version of this book on your computer, copy-and-paste the entire code block for this chapter.

No matter how you got the project code into your Arduino software, save your work, using *File* ➜ *Save*.

Make It Go

Before you upload your code to your Arduino, put the toy or your finger on the sensor—because if there's nothing there, the buzzer will start to scream!

Upload the code to your Arduino by clicking the arrow button at the top of the blue window or using *Sketch → Upload*.

Carefully place your object onto the pressure sensor until it sits there without the buzzer going off. Now, if someone comes by and takes it (or if the cat knocks it over), the buzzer will sound!

FIGURE 7-4: Here's a toy jaguar, holding down the pressure sensor. For the photo, I've replaced the buzzer with an LED, because it's hard to see a buzzer work!

FIGURE 7-5: Pick up the jaguar, and the alarm gets powered—again, here using an LED instead of a buzzer for the visual effect.

Fixes

If the buzzer doesn't stop when you place the toy on the sensor, you may need to reduce the amount of force needed to silence the alarm. That's set as the movedValue number in the code.

```
int movedValue = 100; // threshold value to trigger the buzzer
```

Change 100 to a smaller number, like 50, and give it a try. You can also open the Serial Monitor window using *Tools* ➜ *Serial Monitor* to watch the values change with different pressure and help you pick the right threshold.

WHAT'S GOING ON?

The force sensor conducts electricity better when something is pushing against it, so the sensor numbers rise when there's weight or pressure applied to the disk, and they drop to zero (no current) when nothing is pushing on the sensor.

We watch for that drop toward zero with an "if-then" instruction: *If* the sensor value is less than 100, *then* sound the alarm.

Untethering Your Arduino

If you're actually going to use this to monitor an object, chances are you don't want your computer sitting next to your Arduino the whole time. This is where the optional Arduino power supply comes in. Your Arduino will run the current program in its memory (the last one uploaded) as long as it has power—even if you disconnect your computer.

But since your computer also provides power to the Arduino, you need power from another source. Hence the optional power supply. If you don't have one, there are links to places you can purchase them at http://keefe.cc/stuff-alarm.

TAKING IT FURTHER

Pressure-sensitive Plastic

I'm fascinated by fabrics and materials you can use for sensing. There's material called Velostat that responds to changes in pressure the same way as our pressure sensor. It's flexible and feels like a thick trash bag.

You can cut out much bigger sections of Velostat and place it under your Earth-saving object for a better effect.

Resources for how to do that, and links to places you can buy Velostat, are at http://keefe.cc/stuff-alarm.

CODE CORNER

Pretty much everything in this project's code is something we've already talked about in "Code Corner," so here's something fun to learn about:

Equal Doesn't Equal Equal-Equal

Heh. Say that ten times quickly!

Here's the story. We've been using the equals sign = to assign a value to a variable. Like, "Set buzzerPin equal to 13":

```
int buzzerPin = 13;
```

But what if you want to use an "if-then" statement like this: "If the sensor value is equal to 100, turn on the buzzer." In this case, we're using "equal" as a comparison, not an assignment. See? And that's very different for a computer. So for comparisons, we use equal-equal == instead.

```
if (sensorValue == 100){
  digitalWrite(buzzerPin, HIGH);
}
```

This is a key point, and it's super common to mess up. It's especially tricky because in many computer languages, such as Arduino's, accidentally putting an assignment (one equals sign) instead of a comparison (two equals signs) into an "if-then" statement doesn't cause errors. Worse, it's considered a "True" statement! So the code inside your "if-then" statement *always* runs. And that's probably not what you want.

Electric Candle

How might we make a LED candle you can actually blow out? That was a question my daughter and I had one evening, so we got online and started exploring different ways to sense wind. Along the way, we discovered a pretty nifty sensor that does exactly what we needed. That sensor is the center of this project.

CONCEPTS: SENSING AIR MOVEMENT AND WIND

Weather stations use sets of spinning cups, called *anemometers*, to measure wind. Another way to detect air movement might be to hang two strips of tinfoil next to each other and detect when they touch, completing an electrical circuit—though that might also detect a curious cat.

There's another method, which works on the same principle you use when your pizza slice is too hot to eat and you blow on it to cool it down.

Ingredients

- **1 Arduino Uno**
- **1 Arduino USB cable**
- **1 breadboard**
- **1 push button**
- **1 10k-ohm (10kΩ) resistor, which has a brown-black-orange stripe pattern**
- **1 LED**
- **10 jumper wires**
- **Your computer**

Special items

- **1 Modern Device wind sensor**
- **Soldering iron**
- **Solder**

Optional Items

- **1 Arduino power supply**

In the same way, a wire warmed up by a little electricity will cool when air blows across it—and it's possible to measure that temperature change to sense the wind!

These are called "hot-wire" wind detectors, and there's one that plays nicely with an Arduino.

Almost all of the ingredients you'll need come in typical Arduino starter kits, with the notable exception of the wind sensor. For this project, you'll need one. You can search the Internet for "Modern Device Wind Sensor" or get the link at http://keefe.cc/electric-candle.

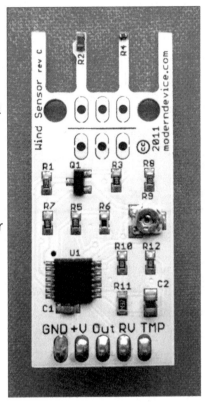

STEPS

The wind sensor requires some assembly. You need to solder the row of pins to the sensor so you can then stick them into the breadboard. This little row of pins is known as the "header."

FIGURE 8-1: The Modern Device wind sensor

Wait, Soldering? Scary!

Not at all!

Soldering is not difficult, but you do need a soldering iron, some solder, and a little lesson.

If you don't have a soldering iron or solder, you can get them at a local hardware or hobby store, and of course from an online seller. Links to some of those are also at http://keefe.cc/electric-candle.

Learning to solder is empowering, awesome, and easy. The best tutorial I know of is at Adafruit. Here's an easy link to it: http://keefe.cc/soldering.

This project is a great time to learn, because you need to solder just five spots: the five pins that make the "header" to stick into the breadboard. (The header pins come with the sensor board.)

The hardest part of soldering is holding everything together while you actually apply the solder to the pins. But in this case, your breadboard can be your assistant.

Just push the long ends of the header pins into the breadboard and place the holes at the end of the wind sensor onto the short ends. To keep the board level, place a coin between the breadboard and the other end of the sensor board.

Then solder away, making sure none of the solder from one pin touches solder from another pin.

Once you have the header attached, you're good to go.

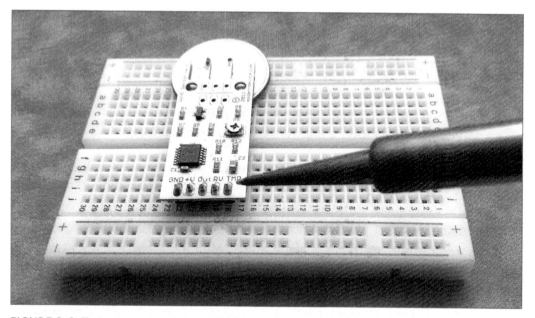

FIGURE 8-2: To free your hands for soldering, push the long ends of the header pins into the breadboard and place the five holes of the wind sensor on top of the short ends. Use a coin to keep the wind sensor level.

Wire Up the Parts

Here's the wiring diagram for this project:

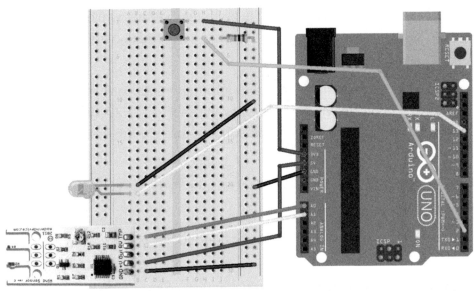

fritzing

FIGURE 8-3: The wiring diagram for the "Electric Candle" project, with the Modern Device wind sensor on the breadboard at the bottom left

- ▸ Insert the button at the top of the breadboard so it straddles the center canal of the breadboard and two of its legs are in Row 1.

- ▸ Grab the resistor, and put one of its legs into the breadboard's Column *j* at Row 3. Take special note here: This leg should be in the same row as a leg of the button. But some buttons are different sizes, and if yours has a leg in a different row, put this resistor leg in *that* row instead. The resistor leg should be in the same row as the button leg (but also not in Row 1).

- ▸ Put the other leg of the resistor in any hole along the blue "–" rail.

- ▸ Grab the LED and insert its shorter leg into the breadboard's Column *a* at Row 20.

▶ Insert the LED's longer leg one hole below, in Column *a* at Row 21.

▶ At the point where the LED legs meet the breadboard, carefully bend the LED at a right angle so it's basically "lying down" on the breadboard.

▶ Insert the 5 pins of the wind sensor into the bottom five rows of the *a* column on the breadboard, so they're in rows 26 through 30. The rest of the wind sensor should extend off of the left side of the board, and the sensor's GND pin should be in Column *a* at Row 30.

Things should now look like this:

FIGURE 8-4: The parts for the "Electric Candle" project, on the breadboard and ready for jumper wires

Time for the jumper wires! Each step below is for one jumper wire, with the connections for both ends. The colors really don't matter, but I'll mention them as they match the illustration.

- ▶ Insert one end of a (red) jumper wire into the Arduino's 3.3V pin, and the other end into the breadboard's Column h at Row 1.

- ▶ Insert one end of a (green) jumper wire into the Arduino's 2 pin along the "Digital" row, and the other end into the breadboard's Column h at Row 3. Note that this end should share a row with one of the button legs and one of the resistor legs. If your button leg is in a different row, use that row instead.

- ▶ Insert one end of a (black) jumper wire into one of the Arduino's GND pins, and the other end into any hole along the breadboard's blue "–" rail on the right—the same rail as one leg of the resistor. This will be our "ground rail," which will provide ground to all of the items on the breadboard.

- ▶ Insert one end of (another black) jumper wire into the same blue "–" rail on the breadboard, and the other end into the breadboard's Column b at Row 20. This is the same row as the short leg of the LED, providing ground, or the negative "–" side of the circuit, to the LED.

- ▶ Insert one end of a (yellow) jumper wire into the Arduino's Pin 13, and the other end in the breadboard's Column b at Row 21—the same row as the longer leg of the LED.

- ▶ Insert one end of a (red) jumper wire into the Arduino's 5V pin, and the other end into the breadboard's Column c at Row 29. Notice that that's the same row as the sensor's +V pin. This wire provides the power to the sensor.

- ▶ Insert one end of an (orange) jumper wire into the Arduino's Pin A0, and the other end in the breadboard's Column c at row 26. This is the same row as the sensor's TMP pin.

▶ Insert one end of a (yellow) jumper wire into the Arduino's Pin A1, and the other end into the breadboard's Column c at Row 27. This shares a row with the sensor's RV pin.

▶ Finally, insert one end of a (black) jumper wire into the breadboard's Column c at Row 30, and the other end into any hole in the blue "−" rail on the right side of the breadboard—our "ground" rail.

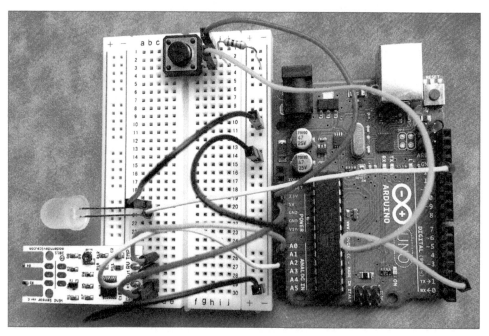

FIGURE 8-5: A photo of the "Electric Candle" project, assembled

WHERE'S THE RESISTOR?

You may have noticed that in almost every other project, we paired the sensors with a resistor. In this project, we do that again for the button (which is really a kind of sensor), but not with the wind sensor. The reason? We've graduated to more sophisticated sensor boards that have the sensor itself but also some additional electronics—those little parts on the wind sensor board—that manage the power and send back to the Arduino just the information we need. To the extent resistors are needed, they're already on board. (Literally!)

Load Up the Code

Get the project code in one of these now-familiar three ways:

FROM THE WEB

▶ Use a web browser to visit http://keefe.cc/electric-candle.

▶ Follow the same instructions as in previous chapters.

FROM THE BUNDLE

Don't have the bundle yet? Check out the instructions in Appendix B. Then . . .

▶ Find your `family-projects-sketches-master` folder, and double-click it to open it.

▶ Click on the `electric_candle` folder.

▶ Inside, you'll see `electric_candle.ino`. It'll have a blue Arduino icon.

▶ To use it, just double-click the file name and it should open in your Arduino software.

FROM THE BACK OF THIS BOOK

If you're reading this book on your computer, you may be able to copy and paste the code from the back of the book into your Arduino software. If that's the case:

▶ On your computer, start up your Arduino software and start a new sketch from the menu bar: *File ➜ New*.

▶ Delete what's in the window that pops up.

▶ Go to Appendix B and highlight the entire code block for this chapter.

▶ Copy the code using *Edit ➜ Copy* from the menu.

- ▶ Switch back to the Arduino software and click into the blank window.

- ▶ Paste the code you copied from Appendix B using *Edit* ➜ *Paste* from the menu.

No matter how you got the project code into your Arduino software, be sure to save your work, using *File* ➜ *Save*.

Make It Go

With the LED glowing, give a good puff across the top of the sensor. The light should go out! Press the button on the breadboard to relight it.

To free your candle from the computer's tether, you can plug an optional power adapter into your Arduino.

You could also power the project with a 9-volt battery. There are some battery holders that have an Arduino adapter *and* a little switch. That's important because even when the candle is out, your Arduino is still running. So power it down completely with the switch (or by unplugging the battery) or you'll run out of juice.

Fixes

The code should work to detect a good puff. If you want to make it more or less sensitive, adjust the wind speed that triggers the dousing by changing the number in the following line from 6 to something else:

```
if (WindSpeed_MPH > 6) {
```

WHAT'S GOING ON?

As air moves across the sensor's "hot wire" (it's not that hot), the wire cools and its conductivity changes. The other electronics on the board detect this change and turns it into values the Arduino can read.

When the values hit a threshold, we douse the light . . . and wait for someone to push the button to relight it.

CODE CORNER

You're a Star

There is some serious math going on in the code for this project. You don't need to know all of what's happening, of course, but I thought it would be a good time to point out some basic math symbols.

You probably gather that + is used for addition and - is used for subtraction.

But what about *? That's multiplication. So 2 * 3 equals 6.

And / is for division. Which means 6 / 3 is 2.

In this code, there's even a pow. That's not comic-book slang for a punch; it's "power," as in 10 to the power of 2. You might know that as 10^2. In Arduino, that's written pow(10, 2). Either way, it's 100!

Fun Functions

If you take a peek at the code for this project, you'll see our old friends void setup() and void loop(). But scroll down a bit and you'll see they're joined by a couple of new sections: void douseCandle() and void lightCandle(). What's going on there?

These are two *functions* I've created to perform a specific task in the code. Basically, I've added two commands to the existing Arduino vocabulary, joining existing commands such as analogRead() and digitalWrite().

Up in the void loop() section, I "call" these functions in a couple of different spots. One of them looks like this:

```
if (WindSpeed_MPH > 6) {
    douseCandle();
}
```

When the program sees `douseCandle()`, it goes and looks for the function I've made, which is written like this . . .

```
void douseCandle() {

  // turn LED off
  digitalWrite(led, LOW);

}
```

When the function is called, the program runs the code between the function's brackets { }—which sets the LED pin to "LOW."

Functions are super useful. For one, they let you run the same block of code in many different spots. You can create a function for that code, and then just call the function whenever you need it.

That way, you don't repeat yourself—which is something coders try to abide by. They even have a name for it: "DRY" code (for Don't Repeat Yourself).

TAKING IT FURTHER

We're interested in whether there is *any* significant air movement across the sensor. But this little device can actually provide good data on *how fast* that air is moving—an actual wind speed detector.

To be accurate, though, the sensor needs its own source of power, separate from the Arduino. That's because the power running through the Arduino can fluctuate slightly, and that will affect the precise measurements of the sensor. Links to more information on how to wire up a separate power supply are at http://keefe.cc/electric-candle.

Invisible Ruler

Maybe you know how bats "see" with sound: emitting high-pitched clicks and listening for how those sounds bounce off things. It's called "echolocation" because the bats literally locate things using echoes. We can make an Arduino do the same thing!

CONCEPTS: DISTANCE SENSING

Detecting objects around us is key to navigating our world, and can be important for smart objects, too. Is the cat nearby? Am I about to bump into a wall?

To give our Arduino "eyes," we'll use a Ping sensor, which, just like a bat, emits high-pitched sounds and listens for them to return. By timing how long it takes the sound to make that super-quick round-trip journey, we can sense when an object is near the sensor *and* measure how far away it is.

As an added bonus, the Ping sensor looks like two little eyes—which can be great if you are building a robot.

Ingredients

- **1 Arduino**
- **1 Arduino USB cable**
- **1 breadboard**
- **3 jumper cables**
- **Your computer**

Special parts

- **1 Ping sensor**

Optional parts

- **1 LED**

The key part here is, of course, the Ping sensor. A quick Internet search will lead you to one, or you can find links at http://keefe.cc/invisible-ruler.

STEPS

Wire Up the Parts

The pins on the Ping sensor are set up in a very common pattern for sensor parts:

 ▶ One pin for power, or positive "+" side of the circuit. On the Ping, it's labeled 5V for 5 volts of power.

 ▶ One pin for the "ground," or negative "−" side of the circuit, marked GND on the Ping.

 ▶ One pin for the data, providing some measure of what the sensor is sensing. On the Ping, that pin is labeled SIG for "signal."

FIGURE 9-1: The pins on the Parallax Ping distance sensor

Wiring up this project is pretty straightforward:

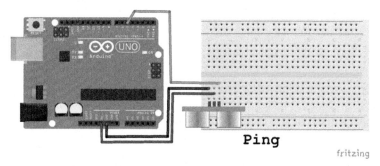

Ping

fritzing

FIGURE 9-2: The wiring diagram for the "Invisible Ruler" project, using a Ping distance sensor. The pins on the Ping are in rows 1 through 3 on the breadboard.

1. Push the Ping sensor pins into the breadboard so that the pins fit into rows 1 through 3, with the Ping's GND pin fitting into breadboard Row 1 and the Ping's SIG pin fitting into breadboard Row 3.

2. Insert a (preferably black) jumper wire into breadboard Row 1, the same row as the Ping's GND pin. Be sure the Ping and the jumper wire are on the *same side* of the canal down the middle of the breadboard.

3. Connect the other end of this (black) jumper wire to one of the Arduino's GND pins.

4. Insert a (preferably red) jumper wire into breadboard Row 2, the same row as the Ping's 5V pin. Again, be sure to stay on the same side of the center canal.

5. Connect the other end of this (red) jumper wire to the Arduino's 5V pin.

6. Insert one end of a third jumper wire (any color!) into breadboard Row 3, the same row as the Ping's SIG pin. Same rule about the canal.

7. Connect the other end of this third jumper wire to Arduino Pin 7.

That's it!

Load Up the Code

Pick your favorite way to get the code:

FROM THE WEB

▶ Use a web browser to visit http://keefe.cc/invisible-ruler.

▶ Follow the same instructions as in previous chapters.

FROM THE BUNDLE

Once you get your bundle . . .

▶ Find your `family-projects-sketches-master` folder, and double-click it to open it.

▶ Click on the `invisible_ruler` folder.

▶ Follow the same instructions as in previous chapters.

FROM THE BACK OF THIS BOOK

Reading this book on your computer? Here are the copy-paste instructions:

▶ On your computer, start up your Arduino software and start a new sketch from the menu bar: *File* ➜ *New*.

▶ Delete what's in the window that pops up.

▶ Highlight the entire code block for this chapter in Appendix B.

▶ Copy the code using *Edit* ➜ *Copy* from the menu.

▶ Switch back to the Arduino software and click into the blank window.

▶ Paste the code you copied from Appendix B using *Edit* ➜ *Paste* from the menu.

No matter how you got the code into your Arduino software, be sure to save your work, using *File* ➜ *Save*.

Make It Go

Upload the code to your Arduino by clicking the arrow button at the top of the blue window or using *Sketch* ➜ *Upload*.

Next, open up the Serial Monitor (*Tools* ➜ *Serial Monitor*) and place your hand in front of the sensor. The numbers should change as you get closer to or farther from the sensor.

The measurements won't be exact, but they're accurate enough to trigger an action if something comes near!

Fixes

If you're seeing nothing or gibberish in the Serial Monitor, be sure the menus at the bottom of that window are set to "9600 baud" and "CR no LF."

Be sure you don't have the power wires reversed: GND on the Ping should connect to GND on the Arduino.

WHAT'S GOING ON?

The Arduino sends out a "HIGH" signal to the Ping, which transmits a super-high frequency sound. The code then waits to see how long it takes to pick up the return signal in microseconds, which are *millionths* of a second! We know that sound takes 74 microseconds to travel an inch (29 microseconds for a centimeter), so we can turn the time gap into the distance the sound covered while we were waiting. Half of that round trip is the distance to the object!

CODE CORNER

Sensing the Speed of Sound

There's a little command in this code that's the key to everything: `pulseIn(pingPin, HIGH)`. This command sits and waits for an input pin to detect something, and lets you know how long it waited. So written as below, duration becomes the number of microseconds the Arduino was waiting:

```
duration = pulseIn(pingPin, HIGH);
```

The crazy (to me) part is that just a couple of lines earlier, we sent a quick "chirp" to the ultrasonic emitter. So `pulseIn` is measuring how long it takes to hear the *echo* off something just a few inches away.

That value, stored in `duration`, is then used to calculate the distance to that object.

Beyond the "Void"

Until now, the separate blocks of code in our sketches have started with void—as in `void setup()` and `void loop()` and `void douseCandle()`. In this project, though, there are some sections, or functions, that begin with `long` instead.

Here's why. We're sending the function a number and expecting something back in return. For example, if the main program sends the function the number 5,800, like so . . .

```
cm = microsecondsToCentimeters(5800);
```

. . . the 5800 gets delivered to the `microsecondsToCentimeters` function . . .

```
long microsecondsToCentimeters(long microseconds)
{
  // The speed of sound is 340 m/s or 29 microseconds per centimeter.
  // The ping travels out and back, so to find the distance of the
  // object we take half of the distance traveled.
  return microseconds / 29 / 2;
}
```

. . . which welcomes that 5,800 as the variable microseconds, divides it by 29 (the / means divide), and divides it further by 2. The result of that math (5,800 / 29 / 2) is 100, so 100 gets *returned* to the main program. So in this case, the variable cm at the very beginning of our example is now equal to 100.

We need to tell the Arduino what kind of value is coming back. That's where long comes in. This particular program could be dealing with some big-time numbers, even more than a billion. The long designation tells the Arduino to save some extra space for that number in its little brain. Numbers designated as long numbers can be between 2,147,483,647 and -2,147,483,648. By comparison, an int number can only get as big as 32,767 (or down to -32,768).

When nothing is expected back from a function, we use the word void instead, as you've been doing up until now.

TAKING IT FURTHER

Take Action

You can tell your Arduino to act on the distance measurement by using some "if-then" code. *If* an object comes within 3 inches, *then* do something (light an LED, ring an alarm, back away).

Let's do that. When something is near, we'll light up an LED. Go ahead and add the optional LED to the Arduino, putting the long leg in Pin 13 and the short leg in GND.

Then add these lines right above the delay(100); line, like this:

```
if (inches < 3) {
    digitalWrite(13, HIGH);
} else {
    digitalWrite(13, LOW);
}
delay(100);
```

The LED should light up whenever your hand is near the sensor!

Get Your Arduino Online

Until now, your smart objects sensed things nearby and communicated with you with blinking lights or numbers on your screen. We can make the Arduino even smarter by letting it surf the Internet! Once online, it can nab weather reports, monitor your environment, and even send emails at the push of a button!

CONCEPTS: CONNECTING TO THE INTERNET WITH A WIFI BOARD

To get online, we need to get your Arduino on to your home wifi network. And to do *that* we need to add some wifi parts to your Arduino. We do that with a new circuit board about the same size as an Arduino that piggybacks on top of your Arduino. In Arduino-speak, these add-on boards are known as "shields." I like to think that shields give Arduinos superpowers.

Ingredients

- **1 Arduino**
- **1 Arduino USB cable**
- **Your computer**
- **A home wifi network and the password for it**

Special items

- **1 SparkFun WiFi Shield (ESP8266)**
- **1 set of Arduino R3 Stackable Headers**
- **Soldering iron**
- **Roll of solder (lead free)**
- **Small piece of tape (any kind will do)**

Optional items

- **1 breadboard**
- **1 set of Break Away Headers**

This particular superpower will be key to the rest of the book; all of the other projects use this board for exciting jobs. So let's take a moment to get one of these little guys up and running.

The SparkFun WiFi Shield is the key link between your Arduino and the Internet. There are links for ordering the wifi board, and the rest of these parts, at http://keefe.cc/arduino-online.

NOTE ABOUT WIFI NETWORKS

The wifi board we're using works nicely on most home wifi networks. But networks at many universities, hotels and cafes have extra steps to join a network—such as pop-up windows asking you for more information. This board won't work on those systems. It has no way to fill out those extra boxes! For possible workarounds, see Appendix A.

STEPS

As with the "Electric Candle" project, we need to solder some headers onto the wifi board. Again, soldering is a whole lot easier than you might think. And there are some great tutorials online, complete with videos. One of my favorite guides is here: http://keefe.cc/soldering.

To help solder the headers on this board, I made a little mounting platform using parts in the "Optional items" section under "Ingredients"—a set of "Break Away Headers." Note that unlike the stackable headers, these are just pins; they don't have any holes.

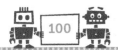

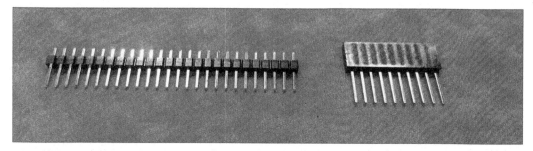

FIGURE 10-1: Two kinds of headers: break-away (left) and stackable (right)

You can break these headers into smaller sections (hence the name). I broke off four groups of five pins and pushed the long ends of each group into the breadboard at rows 1 and 20.

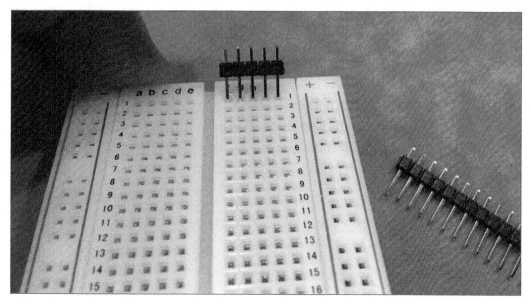

FIGURE 10-2: Inserting the break-away headers into the breadboard

When you're done, it'll look like this:

FIGURE 10-3: Positioning the break-away headers in rows 1 and 20 to make a soldering base

Then I line up the set stackable headers, their pins pointing up, onto the breadboard setup in the same pattern as holes on the wifi shield—if the shield is facing down. That pattern looks like Figure 10-4

Now the pins correspond to the holes on the face-down wifi shield. Note that we're ignoring the holes on the "middle" end of the wifi shield.

Slide the board onto the pins, making sure the side with the electronic components is facing down toward the breadboard.

FIGURE 10-4: Line up the stackable headers onto your breadboard platform.

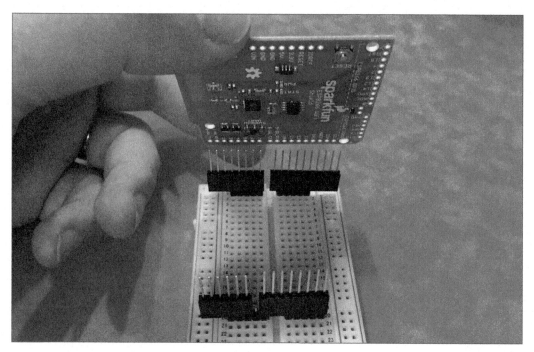

FIGURE 10-5: Slide the wifi board onto the long pins, with the electronics on the board facing the breadboard.

FIGURE 10-6: Arranging the pins this way does the most important job in soldering: holding the board for you so you can focus on the soldering!

Wire Up the Parts

You've accomplished all of the wiring with your awesome soldering job. Nice work.

Before we put the wifi board and the Arduino together, place a small piece of tape on top of the Arduino's USB port—the shiny box you plug the USB cable into. That box actually is connected to "ground," and we want to keep that separated from the metal contacts on the underside of the wifi board so it doesn't accidentally complete a circuit we don't want.

Electrical tape tends to stick to metal better, but any tape will do the trick.

Then carefully press the shield straight down onto your Arduino, making sure all of the pins from the wifi board go into all of the pins on the Arduino.

FIGURE 10-7: The wifi board pressed onto the Arduino. Note that all of the wifi board pins line up with all of the Arduino's holes. To the left, you can also see the blue tape I put on top of the USB port.

Then connect your Arduino to your computer to load up the code.

Load Up the Code

As we've seen before, in the "A Gentle Touch" project, sometimes we need to add to Arduino's knowledge, and that's often true when we add a new part like this. So we'll need the "library" for the wifi shield.

Remember that a "library" isn't a comprehensive as it sounds. It is a little bit of code containing commands for a particular task. Here's how to get the library for the wifi board.

▶ Download the "wifi" library by putting this link into an Internet browser: http://bit.ly/sparkfun-wifi.

▶ That should download a compressed file with a ridiculously long name called `SparkFun_ESP8266_AT_ArduinoLibrary-master.zip`. It will probably end up in your "Downloads" folder. Don't unzip it!

▶ At this point, if you see a folder and not a ".zip" file, your browser may have "helpfully" unzipped the file and deleted the original. Check out the "Was Your File Politely Deleted?" note in the "A Gentle Touch" project.

▶ Start your Arduino software (if it's not running already).

▶ From the menu bar, select *Sketch* ➜ *Include Library* ➜ *Add .ZIP Library* . . .

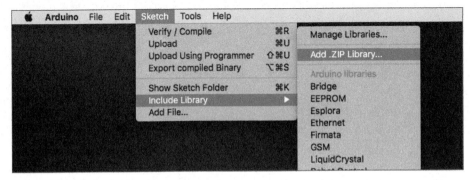

FIGURE 10-8: Adding a downloaded library on a Mac (it's in the same spot in the Windows version)

▶ Navigate to the place your browser saved `SparkFun_ESP8266_AT_ArduinoLibrary-master.zip` (it's very likely in your "Downloads" folder).

▶ Select `SparkFun_ESP8266_AT_Arduino_Library-master.zip` and click the "Choose" button.

▶ Great! Now quit your Arduino software and restart it.

Fortunately, the library you just installed also comes with the code we need to get up and running. Start from the Arduino software menu bar and select *File* ➜ *Examples* ➜ *SparkFun ESP8266 AT Library* ➜ *ESP8266_Shield_Demo*.

FIGURE 10-9: Finding the Demo Sketch under *File* ➜ *Examples*. Shown on a Mac, but it'll be in the same spot no matter your computer.

To make this code work, though, you need to add some information about your own wifi network. Look for this section of the code:

```
///////////////////////////////
// WiFi Network Definitions //
///////////////////////////////
// Replace these two character strings with the name and
// password of your WiFi network.
const char mySSID[] = "yourSSIDhere";
const char myPSK[] = "yourPWDhere";
```

Replace the `yourSSIDhere` with the name of your wifi network, such as "Keefe Family WiFi," and replace the `yourPWDhere` with the password for your network. Be sure to keep the double quotes around each.

Save your sketch using *File* ➜ *Save*. (You'll be prompted to save it with a new name.)

Before you upload your sketch to the Arduino, look for the little switch on the surface of the wifi board marked "UART." If it's not already, slide it to "SW."

Now upload your sketch using *Sketch* ➜ *Upload* or the arrow button at the top of the window.

Make It Go

It's the Moment of Truth!

Open the Serial Monitor window with *Tools* ➡ *Serial Monitor*. You should see

```
Press any key to begin.
```

That's not quite true. You need to put your cursor in the input box at the top of the Serial Monitor window and then press Return or Enter to begin. Do this any time the program says "Press any key."

If everything is in order, you'll see

```
ESP8266 Shield Present
Mode set to station
Connected to: [Your wifi network name]
My IP: 192.168.1.2
Press any key to connect client.
. . . (and a whole bunch of other text) . . .
<body>
<div>
    <h1>Example Domain</h1>
    <p>This domain is established to be used for illustrative examples in
    documents. You may use this domain in examples without prior
    coordination or asking for permission.</p>
    <p><a href="http://www.iana.org/domains/example">More information. . .
</a></p>
</div>
</body>
</html>
0,CLOSED
Press any key to test server.
```

That text scrolling by actually lives on the Internet, on a site called "example. com"—so your Arduino is connected to the Internet! You are ready for the next chapters.

If that didn't happen, check out the "Fixes" below.

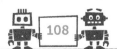

Fixes

If you're having trouble getting your Arduino on the Internet, try these steps:

- Generally, while playing with this board, there are three things I try when I need to jump-start the Arduino, in this order:

  - Close and reopen the Serial Monitor window. This restarts the code on your Arduino.

  - Press the reset button on the wifi board. This also restarts the code.

  - Unplug the Arduino from your computer, wait five seconds, and plug it in again.

- Try those steps if you get `Press any key to begin` but didn't get to see all the scrolling text after you pressed `Return`. Sometimes it takes a couple of attempts to get online.

- Also try those steps if the board doesn't connect to either your wifi or to the Internet right away. Check that your wifi network name and password are entered correctly in the right spot in the code.

- If you have trouble uploading your code to your Arduino—and getting a lot of orange text in your lower Arduino window instead—be sure the little switch on the wifi board is switched to "SW," not "HW," and also double-check that your "Arduino Uno" selections are set correctly under *Tools* ➜ *Board* and *Tools* ➜ *Port*.

- If you're getting random characters in your Serial Monitor window, be sure the drop-down menus at the bottom of that window are set to "Both NL & CR" and "9600 baud."

- On occasion, I've had the entire board just "hang" and not respond to anything. Unplugging the Arduino's USB cable from the computer, waiting a couple of seconds, and plugging it back in has done the trick.

▸ The blue light on the surface of the wifi board provides a little extra information about what's going on:

- Regular flashing means the board is not connected to the wifi network.

- Fast flashing means it's logging into the wifi network.

- Steady means it's connected!

WHAT'S GOING ON?

Just like a laptop computer, the Arduino now has wifi capabilities, which gives it Internet capabilities. The steps it is taking are, essentially:

Arduino ➔ WiFi board ➔ WiFi network ➔ The Internet ➔ example.com

Once it gets there, example.com sends a web page back. Viewed by the Arduino, it's not all nice and pretty, like when you go to http://example.com in a browser:

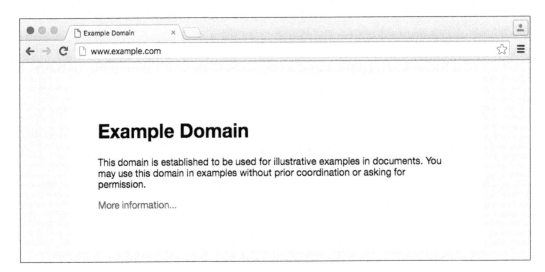

FIGURE 10-10: The website "example.com" in a browser

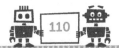

Instead, you're seeing the underlying text, in Hypertext Markup Language (HTML) that your browser turns pretty. Still, you're seeing the same web page.

CODE CORNER

A Quick Tour

Okay, there's a lot going on in this sketch, and you've already done a lot. But here's a quick guide to what's happening:

1. In the beginning of the program, there are a bunch of variables to make the program work. Some variables you customized with your own information.

2. The `void setup()` section runs a whole lot of functions, which are used to get the wifi board up and running.

3. The `void loop()` has just one function inside: `serverDemo()`.

4. Most of the code here is defining the functions called in `void setup()` and `void loop()`.

String Things

In several places you'll see mention of a `string`. Strings are special things in coding that represent words, such as "Hello." The computer, an Arduino in our case, knows to handle them as a *string* of characters—letters, numbers, or symbols—that should be used together.

Strings are often contained in " quote marks, which distinguish them from variables, so string `"hello"` is different from variable `hello`.

The quote marks also let the computer know when numbers should be treated like words, not something you could add or subtract. This comes up a lot with U.S. zip codes, where you might want the string `"02134"` instead of the integer `2134`.

So when you see `String`, think "characters" or "words."

TAKING IT FURTHER

Other Websites

It might seem natural to test out other sites like "google.com" by adjusting the code, and you are absolutely right. Trouble is, that clean, basic Google home page actually has, at last check, 267 lines of code behind it—and many of them are loooong lines! All of that text quickly overwhelms your little Arduino.

In the next chapter, we'll make some precision requests of sites that will get us little bursts of useful information.

What's That "Test Server" Thing?

You may have noticed that the program actually continues beyond where I stopped, with

```
Press any key to test server.
```

Beyond this point, the wifi board actually switches from *looking* at websites to *being* a website—at least within your own home. If you'd like to see how, press Return or Enter here. It should respond with something like

```
Server started! Go to 192.168.1.2
```

The line of numbers may be different for you, but whatever they are, copy the lot of them (in my case 192.168.1.2), switch to a web browser on your computer, and paste them into the location bar at the top of the browser, where you'd normally put a website address:

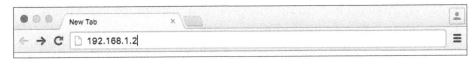

FIGURE 10-11: Put the numbers in the spot you usually type a web address.

Hit return, and you should see a bunch of numbers like

```
A0: 339
A1: 332
A2: 339
A3: 335
A4: 368
A5: 375
```

The super-cool thing? Those numbers are actually readings from the Arduino's six analog pins—A0 through A5! They're pretty random at the moment, because nothing is attached to those pins. But add a sensor circuit, and those values would change with the sensor.

Do I Need an Umbrella Today?

There's a ton of information on the Internet—and some of it is even useful! Now that your Arduino is on the 'net, let's use it to get good data, such as whether you should take an umbrella with you when you leave the house.

CONCEPTS: READING USEFUL INFORMATION ONLINE, USING AN "API"

Yes, people put online lots of cat videos and pictures of pugtatoes (Google it). But some folks also post useful data online, and they make it extra useful by posting it in a way that's easy for computers to read.

This computer-friendly format is called an API, or "another person's information." OK, that's not actually what "API" stands for, but it's far more descriptive than the real definition, "application programming interface."

An API lets you ask for something specific—such as the weather forecast for your area—and get a nicely packaged answer. Let's use an Arduino to make that request and read the answer.

Ingredients

- 1 Arduino
- 1 Arduino USB cable
- 1 LED
- Your computer
- A home wifi network and the password for it
- A web browser

Special items

- Your assembled SparkFun WiFi Shield (ESP8266) from the previous chapter

We'll get your Arduino on to your wifi network, as we did in the previous chapter, "Get Your Arduino Online." If you skipped ahead, and don't have the wifi card set up, head back a few pages for the steps to get started.

STEPS

Wire Up the Parts

Attach the wifi shield you assembled in the previous chapter to the top of the Arduino, being careful to make sure all of the metal pins go into the corresponding holes.

FIGURE 11-1: The DIY sandwich made of an Arduino and the wifi card, topped with an LED (do not eat)

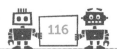

- ▸ Connect the square-ish end of your Arduino USB cable to your Arduino.

- ▸ Connect the flat-ish end of your Arduino USB cable to your computer.

- ▸ Insert the longer leg of the LED into Pin 13 on the wifi board.

- ▸ Insert the shorter leg of the LED into the hole right next door marked GND.

You are ready to go!

Load Up the Code

Pick your favorite way to get your project code:

FROM THE WEB

- ▸ Use a web browser to visit http://keefe.cc/umbrella-today.

- ▸ Follow the same instructions as in previous chapters.

FROM THE BUNDLE

No bundle? Learn how to get it. Then:

- ▸ Find your family-projects-sketches-master folder, and double-click it to open it.

- ▸ Click on the umbrella_today folder.

- ▸ Follow the same instructions as in previous chapters.

FROM THE BACK OF THIS BOOK

And the good ol' copy-paste method, for folks reading this book on a computer:

- ▸ Highlight the entire code block for this chapter printed in Appendix B.

- ▸ Follow the same instructions as in previous chapters.

No matter how you got the project code into your Arduino software, be sure to save your work, using *File → Save*.

Now we need to hop on the Internet to get a little more information that will allow us to communicate with the weather-data site.

Get a Free OpenWeatherMap Key

We'll be using a free service called OpenWeatherMap, which provides weather information around the world. And, fortunately for us, they provide it through an API.

To use the OpenWeatherMap API, we need a key. Keys let data services put some controls on how you use their data, and how often you access their computers. OpenWeatherMap's free plan limits you to 60 requests a minute. Which is plenty for us. Unless you mis-program your Arduino, it's unlikely you'll hit that limit—and the weather doesn't change that often anyway.

To get a key, go to this web address—http://openweathermap.org/appid—and click the "Sign Up" button.

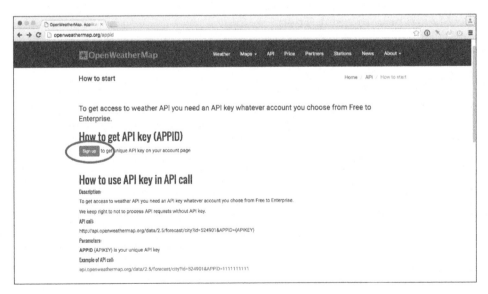

FIGURE 11-2: Sign-up button at OpenWeatherMap

Pick a username, enter your email address, make a password, and check the box if you agree with the terms of service and privacy policy. Then click the "Create Account" button.

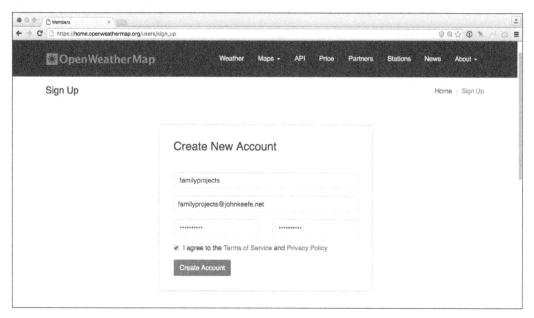

FIGURE 11-3: Creating an account at OpenWeatherMap

You'll then see your shiny new API key!

The key is actually a line of 30 or so letters and numbers that look like a cat walked on your keyboard. Like

```
abc234f53dd994c5445900d00942340c
```

(That key won't work, by the way. My cat made it up.)

Generally, you shouldn't share API keys with anyone else—because you don't want someone to use the service pretending they are you. This is especially important if it's a service you're paying for (which you are not in this case).

▶ Highlight the key with your mouse, making sure to capture all of the characters.

- ▶ Copy the key using using *Edit* ➔ *Copy*. (On a Mac, you can use the `command-c` keyboard shortcut. On Windows and Linux computers, it's `control-c`.)

- ▶ Switch to the Arduino software and the window where the code for this project is visible.

- ▶ Look near the top of the program for these lines:

```
//////////////////////////
// Weather Variables //
//////////////////////////
const String apikey = "YourAPIKeyGoesHere";
const String latitude = "YourLatitudeGoesHere";
const String longitude = "YourLongitudeGoesHere";
```

- ▶ Delete `YourAPIKeyGoesHere` and paste your API key there using *Edit* ➔ *Paste*. (On a Mac, you can use the `command-v` keyboard shortcut. On Windows and Linux computers it's `control-v`.) Here's what my fake key would look like:

```
const String apikey = "abc234f53dd994c5445900d00942340c";
```

Be sure the key sits between the quote marks.

- ▶ Save your changes with *File* ➔ *Save*.

You're ready for the next step!

Figure Out Where You Are

When your Arduino asks OpenStreetMap for the weather where you live, it will need to mention where you live, of course. It could use human words, like "Springfield, Illinois," but translating words into computer-speak can lead to misunderstandings. So we'll identify your town with two numbers: its latitude and longitude.

Don't know your location's latitude and longitude? With the Internet browser on your computer, go to LatLong.net and enter your town's name. (If you don't live in a town, you can use your address.)

FIGURE 11-4: Getting the coordinates for your hometown

So for Springfield, Illinois, the latitude is 39.781721 and the longitude is -89.650148.

Depending on your search, you may get a map pointing to spot that's not your house. That's fine—but just zoom out to make sure the point is, indeed, somewhere in your town (there are a lot of Springfields!).

Let's put that information into your Arduino program:

▶ Highlight the latitude, making sure to capture all of the digits.

▶ Copy the digits using *Edit* ➜ *Copy*.

▶ Switch to the Arduino program and look for these lines near the top:

```
/////////////////////////
// Weather Variables //
/////////////////////////
const String apikey = "abc234f53dd994c5445900d00942340c"
const String latitude = "YourLatitudeGoesHere";
const String longitude = "YourLongitudeGoesHere";
```

▶ Delete the phrase `YourLatitudeGoesHere` (but keep the quotes).

▶ Paste the latitude between the quotes using *Edit ➔ Paste*.

▶ Switch back to your browser.

▶ Highlight the longitude, making sure to capture all of the digits.

▶ Copy the digits using *Edit ➔ Copy*.

▶ Switch to the Arduino program.

▶ Delete the phrase `YourLongitudeGoesHere`.

▶ Paste the longitude between the quotes using *Edit ➔ Paste*.

Note that depending on where you live, the numbers may be negative—and start with a minus sign "−". Be sure to include the minus sign if it's there. If you live in the continental United States, for example, there will be a minus sign before the longitude.

When you're done, it should look something like this—though everything between quotes will be different!

```
/////////////////////////
// Weather Variables //
/////////////////////////
const String apikey = "abc234f53dd994c5445900d00942340c"
const String latitude = "39.781721";
const String longitude = "-89.650148";
```

Almost done!

Enter Your Wifi Information

To get the Arduino on the Internet, you also need to include your wifi network information.

▶ Look for these lines near the top of the Arduino program:

```
/////////////////////////////
// WiFi Network Definitions //
/////////////////////////////
const char mySSID[] = "YourWiFiNetworkNameGoesHere";
const char myPSK[] = "YourWiFiPasswordGoesHere";
```

▶ Replace YourWiFiNetworkNameGoesHere with the name of your home wifi network, keeping the quotes.

▶ Do the same thing with YourWiFiPasswordGoesHere.

Save your changes with *File* ➔ *Save*.

Make It Go

Excellent. Now we're ready to go.

▶ Upload your program to your Arduino using *Sketch* ➔ *Upload* or the arrow button at the top of the Arduino window.

▶ The LED should flash a little when the upload happens.

▶ If it's not already, soon the blue LED on the surface of the wifi shield should shine steadily—without flashing. If it doesn't stay steady, that means it's having trouble connecting to the wifi network. Check out the "Fixes" below.

▶ Once the board is connected, the LED you added will flicker every few seconds. This little "heartbeat" is your sign that the forecast is being checked.

▶ When the Arduino is done checking, that LED will shine bright. If it's steady, no rain is forecast for the next 12–15 hours. If it's flashing like a police car, bring an umbrella!

Next time you need to check, press the "RESET" button on the wifi board, and your Arduino will look up the forecast again. You can keep the whole contraption near your front door, powered by an Arduino power supply.

Fixes

If the LED you added doesn't flicker at all, even when you upload code or press the "RESET" button on the wifi board, check its polarity. Make sure the longer leg is in Pin 13 and the shorter leg is in GND.

If the LED flashes quickly at the start, then stays dark, something is amiss. Be sure all of the information you added to the program is correct, including the wifi network name, password, API key, latitude, and longitude. Note that all of the letters are case sensitive, so copy them exactly as they are.

If the blue LED on the surface of the wifi board isn't steady, that means the board is having trouble connecting to your wifi network. Double-check the network name and password, remembering that they are both case-sensitive.

If you're having trouble getting your Arduino to respond, try these options:

▶ Open the Serial Monitor window with *Tools* ➔ *Serial Monitor*. This restarts the code on your Arduino.

▶ Press the RESET button on the wifi board again. This also restarts the code.

▶ Unplug the Arduino from your computer, wait five seconds, and plug it in again.

WHAT'S GOING ON?

The Arduino asks the OpenWeatherMap folks for the forecast for your location, based on the latitude and longitude. The answer comes back in a long stream of code that includes 40 forecasts: one every three hours for the next five days.

The Arduino looks for a little piece of information, the "weather id," in each the first five forecasts. If the weather id starts with 3 or a 5, rain is forecast for that period. And if any of the first five forecasts indicate rain, the Arduino flashes the LED for a minute. If not, it stays steady.

If you want to *watch* this process in action, just open up the Serial Monitor with *Tools* ➜ *Serial Monitor* and push the RESET button on the wifi board.

CODE CORNER

No Loop!

In this sketch, everything is run out of the `void setup()` section—which calls all of the other functions it needs. That means this whole program runs just once.

This is very un-Arduino! Usually you have code that loops. But in our case, we only need it to happen once. Pressing the RESET button always restarts the program in the Arduino's memory, so that's how we get the program to run again.

Memory Tricks

Your Arduino has a few different sections of memory. One section, the largest, is called the "Flash" memory. It's where your sketch gets stored when you upload it to your Arduino.

There's another, much smaller, section called "SRAM," for static random-access memory, where variables are stored and where words are stored before they are displayed in the Serial Monitor. Like this:

```
Serial.println("Failed to connect to server.");
```

To save on precious SRAM space, this sketch wraps words destined for the Serial Monitor in the F() function, which forces them to be stored in the Flash memory instead:

```
Serial.println(F("Failed to connect to server."));
```

Doing this throughout the sketch for things that don't vary can save valuable space for your variables.

TAKING IT FURTHER

Instead of flashing an LED, what other way might we convey the possibility of rain? If you have LEDs of different colors, can you wire them up so one color indicates rain and another indicates clear?

12

Send Email with a Button

I f making a computer say "Hello, world" is the first step of programming, and making an LED blink is the "Hello, world" of do-it-yourself smart objects, it's quite possible that making an "Internet Button" is the "Hello, world" of things that do things on the Internet. The idea is that you push a button in the physical world and something happens in the Internet world. In our case, we're going to send an email—to you!

CONCEPTS: TRIGGERING THINGS ONLINE, USING IFTTT

For a few projects in a row, we've taught an Arduino to take information *from* the Internet and do something with it. Now we going to send information *to* the Internet, where we'll do something with it.

Ingredients

- 1 Arduino
- 1 Arduino USB cable
- 1 breadboard
- 1 push button
- 1 10k-ohm (10kΩ) resistor, which has a brown-black-orange stripe pattern
- 3 jumper wires
- Your computer
- A home wifi network and the password for it
- A web browser
- An email address

Special Items

- Your assembled Spark-Fun WiFi Shield, from the chapter "Get Your Arduino Online"

Optional Item

- 1 Arduino power supply

Our information is going to be a "trigger" signal, which we'll send with the push of a button.

That signal will go to a free service called IFTTT, which stands for "If This Then That." In our case: *"If* you get a signal from my Arduino, *then* send me an email!"

First, let's put the button and the rest of the ingredients together.

STEPS

Wire Up the Parts

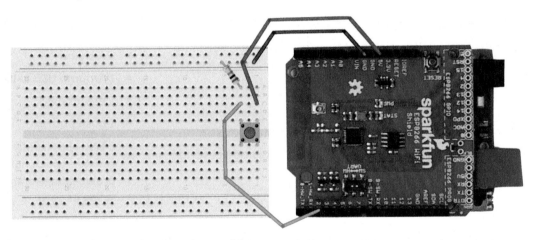

FIGURE 12-1: Wiring diagram for the email button project

▶ Attach the wifi shield you assembled in "Get Your Arduino Online" to the top of the Arduino, being careful to make sure all of the metal pins go into the corresponding holes.

▶ Connect the square-ish end of your Arduino USB cable to your Arduino.

▸ Connect the flat-ish end of your Arduino USB cable to your computer.

▸ Remember that the LED has a longer leg and a shorter leg. Insert the longer leg of the LED into Pin 13 on the wifi board.

▸ Insert the shorter leg of the LED into the hole right next door marked GND.

▸ Attach the push button to the breadboard so that:

   ▪ It straddles the breadboard's center canal.

   ▪ One of the legs is in the hole in Row 1, Column e.

▸ Insert one end of a (preferably red) jumper wire into the Arduino's 5V pin, and the other end into the breadboard's Row 1 at Column c.

▸ Insert one end of another jumper wire (it can be any color; I used blue) into the Arduino's Pin 2, and the other end into the breadboard's Row 3 at Column c.

▸ Insert one end of a (preferably black) jumper wire into any of the Arduino's GND pins, and the other end into the breadboard anywhere along the blue "−" rail on the left side, closest to Column a.

▸ Insert one leg of the resistor into the breadboard's Row 3 at Column a.

▸ Insert the other leg of the resistor into the breadboard anywhere along the same blue "−" rail on the left side.

It should look like this:

FIGURE 12-2: The email button project assembled and ready to send mail

LOAD UP THE CODE

You're surely an expert at getting project code by now, but here are the details for this chapter:

FROM THE WEB

- ▸ Use a web browser to visit http://keefe.cc/email-button.

- ▸ Follow the same instructions as in previous chapters.

FROM THE BUNDLE

▶ Find your `family-projects-sketches-master` folder, and double-click it to open it.

▶ Click on the `email_button` folder.

▶ Follow the same instructions as in previous chapters.

FROM THE BACK OF THIS BOOK

If you're reading this on a computer:

▶ Highlight the entire code block for this chapter in Appendix B.

▶ Follow the same instructions as in previous chapters.

No matter how you got the project code into your Arduino software, be sure to save your work, using *File* ➜ *Save*.

Enter Your WiFi Information

As you did for previous chapters, look for this section and replace `YourWiFi-NetworkNameGoesHere` with your wifi's name and `YourWiFiPasswordGoesHere` with its password:

```
///////////////////////////////
// WiFi Network Definitions //
///////////////////////////////
const char mySSID[] = "YourWiFiNetworkNameGoesHere";
const char myPSK[] = "YourWiFiPasswordGoesHere";
```

Getting Started with IFTTT

All right. The button is ready to send the trigger signal. Now we need to set up IFTTT to *receive* the signal. That requires a free IFTTT account. Here are the steps for making one:

▶ On your computer, open a web browser and go to https://ifttt.com/.

▶ Click the "Sign Up" button in the upper-right corner of the screen.

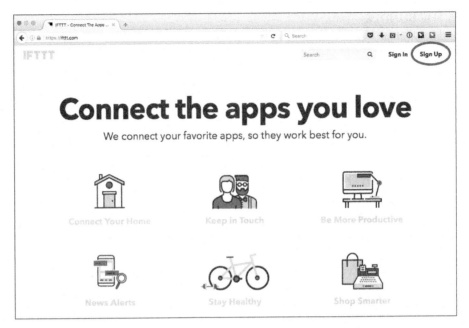

FIGURE 12-3: The IFTTT home page, with the sign-up link highlighted

► Enter your email address.

► Pick a password.

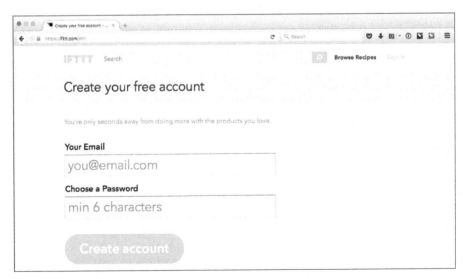

FIGURE 12-4: Pick your email and password for IFTTT. The accounts are free.

Account made! Now, IFTTT takes you on a little tour. Let's get through that.

- ▶ Click the big "This" link.

- ▶ Click the big "That" link.

- ▶ Click the "Continue" button.

- ▶ Click the next "Continue" button.

- ▶ Click three icons that interest you (it doesn't matter which ones at this point).

- ▶ Click "Continue."

Eventually you'll get to a page with this toolbar at the top:

FIGURE 12-5: The IFTTT toolbar, with your name at the far right.

That's where you want to be! There are a bunch of steps below, but they're really straightforward and will only take a couple of minutes.

- ▶ Click on your name in the upper-right corner (if it's not your name, it's part of your email address).

- ▶ Choose "Create."

FIGURE 12-6: The (slightly hidden) IFTTT recipe "Create" link

▶ Click the big "This" link.

▶ You'll see a page that asks you to choose a "Trigger Channel." Either scroll down or use the search box for the "Maker" channel, which has a big "M" logo.

FIGURE 12-7: Searching for the "Maker" channel, which is the big "M" on the left

▶ Click the big "M" logo.

▶ Click the "Connect" button.

▶ Click "Done."

▶ Click "Continue to the Next Step."

▶ You only have one option—"Receive a web request"—so pick that one.

▶ You'll be asked to enter an Event Name. Enter "button_pressed" here.

FIGURE 12-8: Enter "button_pressed" as your IFTTT trigger.

▶ Click "Create Trigger."

▶ Now click the big "That" link.

▶ When you get to "Choose Action Channel," enter "email" into the search box.

▶ Click the plain "Email" icon.

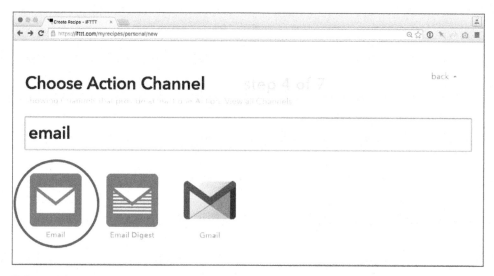

FIGURE 12-9: Searching for the IFTTT "Email" channel. You want the plain Email icon on the left.

► For "Choose an Action," select "Send me an email."

► When you see "Complete Action Fields," just leave everything as it is and click the "Create Action" button.

► At "Create and Connect," press the "Create Recipe" button.

OK! Recipe made. Now we have to get a little bit of information from IFTTT to complete the process.

Get Your Maker Channel Info

We're about to get a little key from IFTTT that we're going to put into our program.

► In your browser, go to http://ifttt.com/maker.

► On this page you'll see a section that says, "Your key is:" with a jumble of letters and numbers.

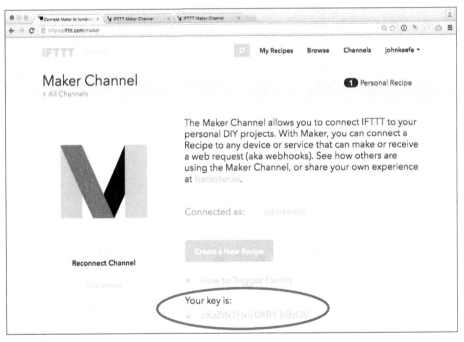

FIGURE 12-10: Finding your "Maker Key" on IFTTT. It's the jumble of characters at the bottom (yours will be different than mine).

- ▶ Copy that jumble by highlighting it, making sure you get all the characters, using *Edit* ➜ *Copy*.

- ▶ Switch to your Arduino software.

- ▶ Near the top of the program, find this section:

```
////////////////////
// IFTTT Variable //
////////////////////
const String maker_key = "YourMakerKeyGoesHere";
```

- ▶ Delete the phrase `YourMakerKeyGoesHere` and replace it with the key from the IFTTT page by pasting the key between the quote marks, using *Edit* ➜ *Paste*.

- ▶ Save your program using *File* ➜ *Save*.

Sweet. You are good to go.

Make It Go

- ▶ Upload the code to your Arduino by pressing the arrow button at the top of the blue window or using *Sketch* ➜ *Upload*.

- ▶ Open the Serial Monitor using *Tools* ➜ *Serial Monitor* just to make sure everything is working. (Opening the Serial Monitor isn't necessary for this program to work, but it lets you see what's happening and helps for troubleshooting.)

- ▶ Wait a full minute. This lets the wifi card get established. The LED will glow during this time.

- ▶ The LED will go dark when everything is ready to send.

- ▶ Push the button!

- ▶ Check your email!

Did you get an email? If not, check out "Fixes" below.

Important note: I've written the program so you have to wait again for a full minute to send another email. This way, if something is amiss, you don't accidentally send yourself hundreds of emails right away (and irritate the servers along the way). The LED will go dark when a minute is over.

Fixes

The best way to troubleshoot this program is to watch two things: the little blue LED on the SparkFun board and the Serial Monitor.

The little blue light needs to be steady and stable. If it isn't, there's a problem staying on the wifi network. One thing that I found helps is to make certain the USB cable is firmly in the computer. Sometimes it slides out even a little, and that seems to cause problems.

The Serial Monitor should look like this:

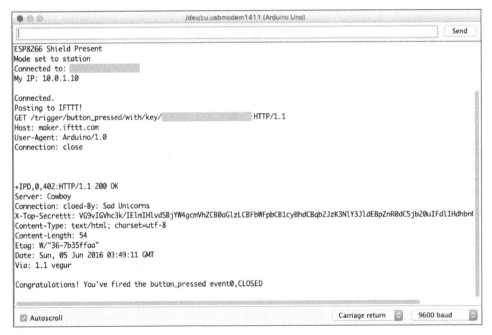

FIGURE 12-11: What things should look like in the Serial Monitor if your button-push is successful

If it doesn't, check out the error solutions below.

LOOPING FOREVER

If you get the "looping forever" error, that means the connection between the Arduino and the wifi board didn't work. Unplug the USB cable from the computer, count to five, and put it back in firmly. There's a chance this was because of a soldering issue, but more likely it is just being flakey.

FAILED TO CONNECT TO SERVER

This I got a lot. The key seems to be that once you press the RESET button, you need to wait for a good minute or two. Waiting allows the board to get connected to the Internet.

Again, make sure the USB cable is firmly in both your computer and your Arduino. That matters.

POSTING TO IFTTT . . . BUT NOTHING

If the script says it's posting to IFTTT, but you see nothing else, be sure your `maker_key` has been copied into your code correctly.

WHAT'S GOING ON?

The Arduino is actually doing something incredibly simple: When you push the button, it essentially visits a web page at IFTTT—and when IFTTT sees the "visit" it triggers the email. You can test this with a web browser, too. Copy this link into your browser . . .

https://maker.ifttt.com/trigger/button_pressed/with/key/YourMakerKeyGoesHere

. . . and before you press `Enter`, change `YourMakerKeyGoesHere` to your unique Maker key, putting it right after the last slash at key/. Then hit `Enter` and you should get an email!

CODE CORNER

The core of this sketch is definitely the `void sendTrigger()` section. That's what actually sends a signal to the IFTTT website. In this section, the communication action happens anywhere you see `client`, such as:

```
client.print(httpHeader);
client.println();
```

With every instance of `client`, the sketch is communicating with the wifi board. Those two lines actually "print" the website information to the wifi board, causing it to visit the IFTTT website and trigger the Maker part of the recipe.

TAKING IT FURTHER

Untethering Your Button

You don't need to keep your Arduino tethered to your computer for this program to work. Simply unplug the USB cable from your Arduino and power it instead with your Arduino power supply. Your program will be active and running a few moments after your Arduino powers up.

Customizing the Email

The email you get is a little robotic, and not particularly friendly.

But you can customize it!

- ▶ Go to IFTTT in your web browser.

- ▶ Click "My Recipes" in the toolbar.

- ▶ Pick your Maker + Email recipe.

- ▶ Change the "Subject" to anything you want, like "Someone pushed the button!"

The event named "button_pressed" occurred on the Maker Channel

IFTTT Action <action@ifttt.com>   Sat, May 21, 2016 at 12:59 PM
To: familyprojects@johnkeefe.net

What: button_pressed
When: May 21, 2016 at 12:59PM
Extra Data: , , ,

IFTTT

Put the internet to work for you.   Turn off or edit this Recipe

FIGURE 12-12: The generic email IFTTT sends

▶ Change the "Body" to anything you want, too.

▶ When you're in one of the two boxes, you'll see a blue bottle icon that gives you the extra values you can include, such as when it happened.

Trigger Something Else

There are dozens of things you can make your button do besides send email. By activating new channels in IFTTT, you can change the "then" part of "If the button is pressed, *then* . . ." Here are some possibilities:

▶ Call your phone and *tell* you(!) the button was pushed (using the Phone Call channel)

▶ Post a tweet on Twitter (using the Twitter channel)

▶ Add a row to a Google spreadsheet (using the Google Drive channel)

▶ Send a notification to your smartphone (using the IF Notifications channel and the IF app on your phone)

In most cases, you'll need to activate the channel, which may include verifying your information (like your phone number) or getting your permission to use another service (like Google).

There are lots of other channels, many attached to brands and companies that sell Internet-connected devices, such as lamps and appliances. If you have one, you can trigger it with your new button!

Online Temperature Tracker

When explorers sailed the seven seas finding new (to them) places, plants, and peoples, they often kept careful logs—a diary noting the day, the time, and what they observed. Arduinos may not be seaworthy, but they are excellent at logging their observations. With this project, we'll get Arduino logging and we'll put that log online.

CONCEPTS: LOGGING AND CHARTING SENSOR DATA ONLINE

Keeping track of observations is super useful. With lots of observations over time, you can see the peaks and valleys in your numbers and discover things you didn't know.

We'll be tracking the temperature inside your house (or wherever you set up your Arduino). If you were to do this by hand, you might set an alarm clock to check a thermometer every hour and then write it down. That's exactly what we'll have your Arduino do—with the added bonus that it'll work even when you are asleep or away!

Ingredients

- 1 Arduino
- 1 Arduino USB cable
- 1 Arduino power supply
- 1 breadboard
- Your computer
- A home wifi network and the password for it
- A web browser

Special items

- Your assembled Spark-Fun WiFi Shield
- 1 thermistor (model TMP36)

Instead of paper, we'll use a free online service called data.sparkfun.com to store the temperatures—which will allow you to see and use the data.

The wifi board is the same one we assembled in the chapter "Get Your Arduino Online." If you've skipped ahead, you'll need to go back to that section before doing this project.

The thermistor comes in almost every Arduino starter kit, and can be also purchased separately online. For links to kits and parts for this project, visit http://keefe.cc/temp-tracker.

STEPS

Wire Up the Parts

The setup for this project almost exactly the same as the one in "Ice, Ice Blinky," but with the SparkFun wifi board mounted on top of the Arduino.

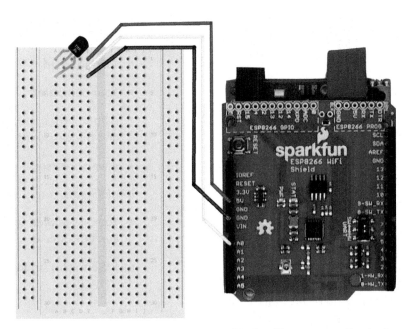

FIGURE 13-1: The wiring diagram for the "Online Temperature Tracker" project, with the SparkFun wifi board atop the Arduino

Here's how to wire it up:

GETTING THE WIFI SHIELD ON TOP

▶ Attach the wifi shield you assembled in "Get Your Arduino Online" to the top of the Arduino, being careful to make sure all of the metal pins go into the corresponding holes.

▶ Connect the square-ish end of your Arduino USB cable to your Arduino.

▶ Connect the flat-ish end of your Arduino USB cable to your computer.

WIRING UP THE THERMISTOR

▶ Notice that the thermistor has three legs, and also that it has both a flat side and a rounded side.

▶ Insert the three legs of the thermistor into in the breadboard's Column *a* at rows 1, 2, and 3 *with the flat side facing the breadboard's printed numbers*. The rounded side will be facing the other direction, toward the

FIGURE 13-2: The position of the temperature sensor, with the flat side facing at the printed numbers along Column *a*

breadboard's center canal. You'll have to separate the legs a little to get them in each of the holes.

▶ Connect one end of a jumper wire (preferably a red one) to the 5V pin on the wifi board and connect the other end to the breadboard's Row 1— let's use Column e.

▶ Connect one end of another jumper wire (any color; I used yellow) to the wifi board's A0 pin, and the other to the breadboard's Row 2 at Column e. Notice that we're staying on the same side of the breadboard's center canal.

▶ Connect one end of a (preferably black) jumper wire to any of the wifi board's GND pins, and the other end of the wire to the breadboard's Row 3 at Column e.

Load Up the Code

Use your favorite code-fetching method one last time, with feeling:

FROM THE WEB

▶ Use a web browser to visit http://keefe.cc/temp-tracker.

▶ Follow the same instructions as in previous chapters.

FROM THE BUNDLE

▶ Find your `family-projects-sketches-master` folder, and double-click it to open it.

▶ Click on the `temp_tracker` folder.

▶ Follow the same instructions as in previous chapters.

FROM THE BACK OF THIS BOOK

▶ Highlight the entire code block for this chapter from Appendix B.

▶ Follow the same instructions as in previous chapters.

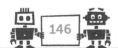

No matter how you got the project code into your Arduino software, be sure to save your work, using *File* ➜ *Save*.

Get a Free Data Account

To store your data online, you need to create a free account at data.sparkfun .com. So let's do that:

- ▶ With an Internet browser, go to http://data.sparkfun.com.

- ▶ Click on the red button that says "Create a free data stream immediately at data.sparkfun.com."

- ▶ Give your data a title, like "Temperature At Our House" (keep in mind things here will be made public).

- ▶ Add a description, like "Temperature sensor in the living room."

- ▶ Choose "public" for your data. This will make it easier to chart your information later.

- ▶ Under "Fields," type temp. This one is important; make sure it's just temp and that it's lower-case.

- ▶ Give it an "alias," which is just a short name you'll use to find your data. Something like karenstemps.

- ▶ The tags and location areas are optional.

Okay! Now you get to a page with a lot of numbers and letters on it, including things like a "Private Key." You'll need all of these, and should have them handy down the road. SparkFun makes keeping track of them really easy with a box at the bottom to "Send yourself a copy of your keys." Let's do that now.

- ▶ Put your email address into the box at the bottom of the page, to make sure you get a copy of everything on this page. Whew!

- ▶ Next, highlight and copy the line of letters and numbers labeled "Public Key" using *Edit* ➜ *Copy* from your browser's menu bar.

▶ Switch to the Arduino software.

▶ Using *Edit* ➜ *Paste*, paste your public key near the top of the program in place of YourPublicKeyHere.

```
/////////////////////////
// SparkFun Data Keys //
/////////////////////////
const String publicKey = "YourPublicKeyHere";
const String privateKey = "YourPrivateKeyHere";
```

▶ Repeat the copy-paste process with your the letters and numbers labeled "Private Key," pasting them into your code in place of YourPrivateKeyHere.

▶ Finally, go back to your web browser and click the link in the "Public URL" box. It'll take you to a mostly blank page. This is your data page, where your data will appear. Bookmark it so you can come back. (It's also in the email you sent yourself.)

Great! We're almost there.

Insert Your WiFi Info

Once again, we need to put your wifi network information into the Arduino program. So switch back to your Arduino software.

▶ Look for these lines near the top of the Arduino program:

```
///////////////////////////////
// WiFi Network Definitions //
///////////////////////////////
const char mySSID[] = "YourWiFiNetworkNameGoesHere";
const char myPSK[] = "YourWiFiPasswordGoesHere";
```

▶ Replace YourWiFiNetworkNameGoesHere with the name of your home wifi network, keeping the quotes.

▶ Do the same with YourWiFiPasswordGoesHere.

▶ Save your work using *File* ➜ *Save*.

Make It Go

▶ Upload the code to your Arduino by pressing the arrow button at the top of the blue window or using *Sketch* ➜ *Upload*.

▶ Open the Serial Monitor using *Tools* ➜ *Serial Monitor* to watch the progress as it works!

On startup, and every two minutes after that, the Arduino will take a reading from the temperature sensor and send it to your page on data.sparkfun.com. Let's see if it's there.

▶ Go back to your web browser and the "data page" you bookmarked (the link is also in the email you sent yourself if you didn't bookmark it).

▶ Use your browser's "reload" or "refresh" button to get a fresh version of the page.

▶ You should see data like this:

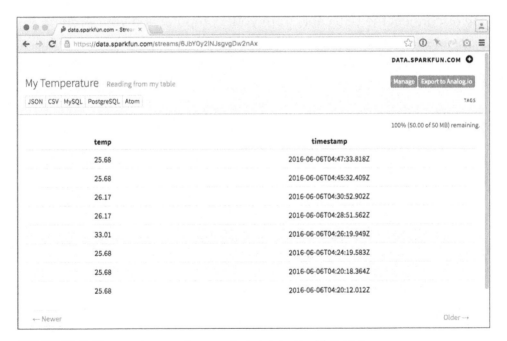

FIGURE 13-3: Temperature readings posted at data.sparkfun.com

If you don't see information like this, check out the "Fixes" section below.

Try holding the thermistor in your hand or touching it to a piece of ice. Wait at least a minute and then refresh the data page again. See if the new values reflect what you did.

If all is well, let's make the check every hour instead of every two minutes.

▶ Back in the Arduino code, find this line and change the 2 to 60:

```
///////////////////
// Sensor Values //
///////////////////
int waitMinutes = 2;
```

▶ Upload the code to your Arduino again by pressing the arrow button at the top of the blue window or using *Sketch* ➜ *Upload*.

Now you're set!

Your temperature logger no longer needs your computer to function, so you can disconnect the USB cable from your computer and your Arduino. To make it operate solo, simply plug in the Arduino power supply. The code will automatically start running again once the Arduino is powered up, and it'll check—and post—the temperature every hour as long as it has power.

Fixes

If you are seeing entries on data.sparkfun.com, but they are all zeros, there are a couple of things to check:

▶ Make sure the column name above the zeros says "temp." That has to match the Arduino code, which is sending a value for "temp." If the top of the column says something else, here's how to fix it:

▪ On your data page, click the "Manage" button at the top right of the page.

▪ When asked, enter your private key for your data, which is in the email you sent yourself (and also in your Arduino code now).

- Click "Edit."

- In the "Fields" section, edit what's there to include one called "temp" (and delete any you're not using, clicking the "x" next to its name).

- Click "Save."

▶ Make sure your thermistor is connected properly. You can check this by opening up the Serial Monitor using *Tools* ➜ *Serial Monitor*. The readings will appear here before getting sent to SparkFun. If they aren't values you'd expect, carefully go through the "Wire Up the Parts" section above.

WHAT'S GOING ON?

Here's the entire chain of events:

▶ The ambient temperature changes the resistance of the thermistor.

▶ Every hour, your Arduino checks the electricity passing through the thermistor, detects the resistance, and converts that into degrees.

▶ It then sends that number over the Internet to data.sparkfun.com, using the same technique you use when you submit the word "Arduino" to a search site.

▶ The SparkFun site adds a timestamp and stores the value online for you.

This all happens super quickly!

CODE CORNER

Pretty Printing

Whether you're printing data to the screen in the Serial Monitor or "printing" data to another server, via the wifi board, how you format your printing can be key.

Throughout this book, you've probably seen `Serial.println()` and `Serial.print()`. Notice the subtle differences—the first one has an extra `ln` after the `print` part. That reads as "print-line," which means print what's in the parentheses, and then make a new line.

So this code . . .

```
Serial.print("The number of legs on an octopus is ");
Serial.println(8);
```

. . . would show up in the Serial Monitor as . . .

```
The number of legs on an octopus is 8
```

Notice the extra space I added after `is` so the 8 wouldn't smash up against it, like `is8`.

Instead of the number 8, I could have used a variable in that second set of parentheses, like so:

```
int legs = 8;
Serial.print("The number of legs on an octopus is ");
Serial.println(legs);
```

which would output the same result.

This is the niftiest way to get variables into lines that you print to the Serial Monitor.

TAKING IT FURTHER

Once you have your data online, now what?

Well, you can download it to your own computer if you wanted, using the "CSV" button. The file you download will open in any spreadsheet program. From there, you can analyze it as you please.

You've also created a kind of API (remember: Another Person's Information or, really, application programming interface). If you right-click the JSON button, you can copy the link location of that button. That Internet address is a live link to your data. Since you made your data public, that link can be used by other programmers to make their own smart objects!

More fun for you, though, may be to *visualize* your data.

Chart Your Data

Since your data is now online, you can use many different services to play with it. One is called Plot.ly, at http://plot.ly, which will allow you to quickly chart your data for free.

- ▸ In your web browser, open a new tab or window.

- ▸ From your data page at data.sparkfun.com, right-click or control-click the the "CSV" button near the top, and choose "Copy Link Address" (or the similar option on your system).

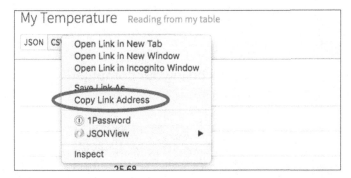

FIGURE 13-4: Right-click the "CSV" button to copy the link to your data.

- ▸ Now go to http://plot.ly and click the "New Chart" button.

- ▸ You'll be asked to log in or sign up. Choose the "Sign Up" tab, enter your details, and pick a password (it's free).

- ▸ You'll get a screen full of stuff, but the data will be missing. Click the "Import Data" link near the top.

FIGURE 13-5: On the Plot.ly page, choose "Import Data."

- ► In the next window, click "By URL."

- ► The link to your data CSV should still be in your computer's memory, so paste it into the box using *Edit* ➜ *Paste*.

FIGURE 13-6: Use the "By URL" option and paste in the address to your data CSV.

- ▶ Press Enter or Return.

- ▶ Under "Chart Type," pick "Line plot."

- ▶ For the "X" box, select "timestamp."

- ▶ For the "Y" box, select "temp."

With those settings, you should see your data visualized!

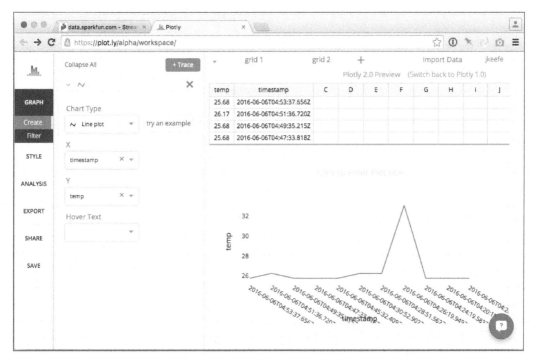

FIGURE 13-7: After setting the Chart Type, the X and the Y values on the left, I get a chart of my data on the right.

Plot.ly has lots of other features you can explore to adjust and share your chart. The entire service is just one example of how you can use your data once it's online and available to play with.

Afterword:

Make More Projects

Congratulations! No matter how many projects you've completed in this book, you are set to build more smart objects.

If you come up with an idea for a smart object, check online to see if anyone has made something similar. Do an Internet search and include the word "arduino," like "automatic cat feeder arduino." Arduino enthusiasts are pretty great about sharing their process and their code. (And there are *lots* of guides to building cat feeders!)

Want to browse possible projects? Here are some good places to start:

▸ Adafruit Arduino Lessons: Take what you learned here and keep going with motors, displays, and more (http://keefe.cc/more-1).

▸ Hackster.io Arduino Projects: Pages and pages of projects contributed by Arduino makers like you (http://keefe.cc/more-2).

▸ The Arduino Playground: A loooong list of projects people have built and posted. Some are super simple, and some are crazy complicated. All are fascinating. Be sure to scroll down to the "Top 40" (http://keefe.cc/more-3).

The more you explore, the more you'll see different flavors of Arduinos beyond the Arduino Uno you've been using. These will include tiny ones, sewable ones, and big-brained ones. Each will have its own quirks and features, but almost all of them can be programmed with the Arduino software you've installed. For guides and tutorials on all kinds of Arduinos, I often turn to the SparkFun and Adafruit websites.

Wherever your explorations take you from here, I'd love to hear about it! Tweet me on Twitter at `@jkeefe` or drop me a note at `familyprojects@johnkeefe.net`.

Happy making!

—John Keefe, Summer 2016

Everything You Need

INGREDIENTS

Each project has a small list of parts you'll need to build the smart object in that chapter. Each chapter also includes a a list of just the parts you need.

If you want to start out with *everything* you need for the entire book, or just want a picture of what's ahead, here's a list of kits that have many of the parts (and often far more), and notes on special parts for different chapters.

All of the parts, and links to the kits, can be found at http://keefe.cc/ family-projects.

Kits

These kits have almost all of the ingredients you'll need for projects in this book.

- ▸ Adafruit Experimentation Kit for Arduino ($$)
- ▸ Adafruit Starter Pack for Arduino ($$)
- ▸ Adafruit Budget Pack for Arduino ($)
- ▸ Make: Getting Started with Arduino Kit, Special Edition ($$)

- ► SparkFun Inventor's Kit for Arduino ($$$$)

- ► SparkFun Inventor's Kit for Arduino – V3.2 ($$$)

Special Parts

For the projects in the last half of the book, starting with "Get Your Arduino Online," your smart objects will be communicating over the Internet, so you'll need this wifi board and its mounting headers:

- ► 1 SparkFun WiFi Shield (ESP8266)

- ► 1 set of Arduino R3 stackable headers

The "Electric Candle" project has a special part—a wind sensor—you won't find in the kits.

- ► 1 Modern Device wind sensor

The "Invisible Ruler" project also has a special part that uses sound to detect distance.

- ► 1 Parallax Ping sensor

Many of the projects suggest an optional power supply, which allows you to untether your Arduino from your computer—and let it run on its own. Arduinos get their power either from the USB port or from the round "barrel" jack next to the Arduino's USB outlet. If your kit doesn't come with a separate power supply, you can buy one that fits the barrel jack or a nifty version that uses your Arduino's USB cable instead.

All of the Parts, Individually

Here's every ingredient used in the book, in order of appearance:

- ► 1 Arduino Uno (Revision 3)

- ► 1 Arduino USB cable

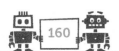

- 1 LED

- 1 breadboard

- 10 jumper wires

- 1 photoresistor

- 2 10k-ohm (10kΩ) static resistors (Stripes: Brown, Black, Orange, Gold)

- 1 static resistor with the highest resistance you have

- 1 photoresistor

- 1 Arduino power supply

- 2 extension jumper wires (male-to-female)

- 1 hollow, translucent toy

- 1 thermistor

- An ice cube

- 1 square of aluminum foil, about the size of a cracker

- 2 pieces of tape

- 1 pressure sensor

- 1 little buzzer

- A small object you cherish :-)

- 1 Modern Device wind sensor

- 1 Parallax Ping sensor

- 1 SparkFun WiFi Shield (ESP8266)

- 1 set of Arduino R3 stackable headers

- 1 set of break-away headers

You can buy the parts individually, by project, or all together, at http://keefe.cc/family-projects.

TOOLS

A couple of the chapters, including "Get Your Arduino Online," require a small amount of soldering. You'll need:

▸ Soldering iron

▸ Solder

. . . and read one of my favorite online guides to soldering at http://keefe.cc/soldering.

WIFI SERVICE

For the last few projects in the book, we get your smart objects online to fetch and send data via the Internet.

The kind of wifi network we use is typical for home wifi setups—at least in the United States: You pick the network from a list of networks in range and, usually, you enter a password into your device for that network.

As long as you know the network name and the password, you should be good.

More problematic wifi networks are those that pop up a window that looks like a web page and asks for information. This could be a username, a hotel room number, or even just a box you need to check to "accept" the terms for using the network. Such networks are often found in hotels, cafes, and universities, and they won't work for this book because the Arduino can't answer the questions or navigate this window.

If you think you're in that situation, you have a few options:

- ▶ Just use the first half of the book. The first eight projects don't need wifi at all. And you'll learn a ton.

- ▶ Work in a location that has wifi without a pop-up window. Most home and business wifi networks should work.

- ▶ See if you can make your cell phone a "hot spot." Check with the cellular provider you use to find out.

- ▶ Buy a "hot spot." These are little boxes that connect to the Internet like a cell phone and create a little wifi network. They typically cost very little, but can have expensive monthly fees.

- ▶ Check with your cable company. Many of them offer subscribers access to their wifi networks around town.

All of the Code

For reference, and for those who wish to program their Arduino by typing the code by hand, I've printed every sketch used in this book below, listed by chapter.

Updates to the code, and copyable code blocks (for those who'd rather not type it out) are also available at http://keefe.cc/family-projects.

Finally, every sketch used in "Family Projects for Smart Objects" is also available as a zipped-up bundle you can download for free. Here's how:

- ▶ Use a web browser to visit http://keefe.cc/sketches.

- ▶ Your browser should start downloading the file.

- ▶ Once it's done, navigate to your "Downloads" folder.

- ▶ If your browser hasn't done it already, unzip the file called `family-projects-sketches-master.zip`, which is usually done just by clicking or double-clicking its icon.

- ▶ You'll see a bunch of "chapter" folders with short names relating to each project in this book. Click on the folder you want, such as `electric_candle`.

▶ Inside, you'll see a file of the same name, ending in .ino (for Arduino), such as electric_candle.ino. It'll have a blue Arduino icon.

▶ To use it, just double-click the file name and it should open in your Arduino software.

HELLO BLINKY WORLD

This program comes with the Arduino software, and is available from the menu bar by choosing *File* ➜ *Examples* ➜ *01.Basics* ➜ *Blink*. The sketch is also available at http://keefe.cc/hello-blinky or in the downloadable zip file as hello_blinky.ino.

```
// the setup function runs once when you press reset or power the board
void setup() {
  // initialize digital pin 13 as an output.
  pinMode(13, OUTPUT);
}
// the loop function runs over and over again forever
void loop() {
  digitalWrite(13, HIGH);   // turn the LED on (HIGH is the voltage level)
  delay(1000);              // wait for a second
  digitalWrite(13, LOW);    // turn the LED off by making the voltage LOW
  delay(1000);              // wait for a second
}
```

A DARK-DETECTING LIGHT

This code comes with the Arduino software, and can be found from the menu bar by choosing *File* ➜ *Examples* ➜ *03.Analog* ➜ *AnalogInput*. The sketch is also available at http://keefe.cc/dark-detector or in the downloadable bundle as dark_detector.ino.

```
/*
  Analog Input
 Demonstrates analog input by reading an analog sensor on analog pin 0 and
 turning on and off a light emitting diode (LED) connected to digital pin 13.
```

```
  The amount of time the LED will be on and off depends on
  the value obtained by analogRead().
  The circuit:
  * Potentiometer attached to analog input 0
  * center pin of the potentiometer to the analog pin
  * one side pin (either one) to ground
  * the other side pin to +5V
  * LED anode (long leg) attached to digital output 13
  * LED cathode (short leg) attached to ground
  * Note: because most Arduinos have a built-in LED attached
  to pin 13 on the board, the LED is optional.
  Created by David Cuartielles
  modified 30 Aug 2011
  By Tom Igoe
  This example code is in the public domain.
  http://www.arduino.cc/en/Tutorial/AnalogInput
   */
int sensorPin = A0;     // select the input pin for the potentiometer
int ledPin = 13;        // select the pin for the LED
int sensorValue = 0;    // variable to store the value coming from the sensor
void setup() {
  // declare the ledPin as an OUTPUT:
  pinMode(ledPin, OUTPUT);
}
void loop() {
  // read the value from the sensor:
  sensorValue = analogRead(sensorPin);
  // turn the ledPin on
  digitalWrite(ledPin, HIGH);
  // stop the program for <sensorValue> milliseconds:
  delay(sensorValue);
  // turn the ledPin off:
  digitalWrite(ledPin, LOW);
  // stop the program for for <sensorValue> milliseconds:
  delay(sensorValue);
}
```

NIGHT LIGHT

The sketch is also available at http://keefe.cc/night-light or in the download-able bundle as night_light.ino.

```
/*

 Night Light
 Lights an LED when the room is dark.
 By John Keefe, January 2016
 http://keefe.cc/night-light
 Based on example code created by David Cuartielles
 modified 30 Aug 2011 by Tom Igoe
 http://www.arduino.cc/en/Tutorial/AnalogInput
 This code is in the public domain.

 */
int sensorPin = A0;    // select the input pin for the photoresistor
int ledPin = 13;       // select the pin for the LED
int sensorValue = 0;   // variable to store the value from the sensor
int darkPoint = 0;     // if the sensor value is less than this, the room is "dark"
// This code runs once when Arduino is turned on or reset
void setup() {

  // declare the ledPin as an OUTPUT:
  pinMode(ledPin, OUTPUT);
  // this is needed to use the Serial Monitor
  Serial.begin(9600);

}
// This code loops forever
void loop() {

  // read the value from the sensor:
  sensorValue = analogRead(sensorPin);

  // if the sensorValue is less than the darkPoint
  // power the LED:
  if (sensorValue < darkPoint) {
    digitalWrite(ledPin, HIGH);
  }
  // if the sensorValue is greater than or equal to the
  // darkPoint, don't power the LED:
```

```
  if (sensorValue >= darkPoint) {
    digitalWrite(ledPin, LOW);
  }
  // output the sensor value to the Serial Monitor
  Serial.println(sensorValue);
  // add a little delay to help read the numbers!
  delay(100);
}
```

ICE, ICE, BLINKY

The sketch is also available at http://keefe.cc/ice-blinky or in the download-able bundle as ice_blinky.ino.

```
/*
  Analog Input
Demonstrates analog input by reading an analog sensor on analog pin 0 and
turning on and off a light emitting diode (LED) connected to digital pin 13.
The amount of time the LED will be on and off depends on
the value obtained by analogRead().
The circuit:
* Potentiometer attached to analog input 0
* center pin of the potentiometer to the analog pin
* one side pin (either one) to ground
* the other side pin to +5V
* LED anode (long leg) attached to digital output 13
* LED cathode (short leg) attached to ground
* Note: because most Arduinos have a built-in LED attached
to pin 13 on the board, the LED is optional.
Created by David Cuartielles
modified 30 Aug 2011
By Tom Igoe
This example code is in the public domain.
http://www.arduino.cc/en/Tutorial/AnalogInput
Further Modified by John Keefe May 2016
to add Serial Monitoring

  */
int sensorPin = A0;    // select the input pin for the potentiometer
int ledPin = 13;       // select the pin for the LED
int sensorValue = 0;   // variable to store the value coming from the sensor
void setup() {
```

169

```
    // declare the ledPin as an OUTPUT:
    pinMode(ledPin, OUTPUT);
    Serial.begin(9600);  // < - - Added this line
}
void loop() {
    // read the value from the sensor:
    sensorValue = analogRead(sensorPin);
    // turn the ledPin on
    digitalWrite(ledPin, HIGH);
    // stop the program for <sensorValue> milliseconds:
    delay(sensorValue);
    // turn the ledPin off:
    digitalWrite(ledPin, LOW);
    Serial.println(sensorValue); // < - - New line is here
    // stop the program for for <sensorValue> milliseconds:
    delay(sensorValue);
}
```

A GENTLE TOUCH

This program is also available at http://keefe.cc/gentle-touch or in the downloadable zip file as gentle_touch.ino.

```
/*
 * CapacitiveSense Library Demo Sketch
 * Paul Badger 2008
 * Uses a high value resistor e.g. 10M between send pin and receive pin
 * Resistor effects sensitivity, experiment with values, 50K - 50M. Larger
resistor values yield larger sensor values.
 * Receive pin is the sensor pin - try different amounts of foil/metal on this pin
 */
// Part of the CapactitiveSense library now maintained by Paul Stoffregen
// https://github.com/PaulStoffregen/CapacitiveSensor
// released under the MIT License.
/*
CapacitiveSense.h - Capacitive Sensing Library for 'duino / Wiring
https://github.com/PaulStoffregen/CapacitiveSensor
http://www.pjrc.com/teensy/td_libs_CapacitiveSensor.html
http://playground.arduino.cc/Main/CapacitiveSensor
Copyright (c) 2009 Paul Badger
Updates for other hardware by Paul Stoffregen, 2010-2016
vim: set ts=4:
```

```
*/
// Modified here by John Keefe May 2016
#include <CapacitiveSensor.h>
CapacitiveSensor   cs_4_6 = CapacitiveSensor(4,6);
int LED = 13;
long sensor_reading;
void setup()
{
   pinMode(LED, OUTPUT);
   Serial.begin(9600);
}
void loop()
{
    sensor_reading =  cs_4_6.capacitiveSensor(30);

    Serial.println(sensor_reading);

    if (sensor_reading > 100) {
      digitalWrite(LED, HIGH);
    } else {
      digitalWrite(LED, LOW);
    }
    delay(10);
}
```

MOVED STUFF ALARM

This program is also available at http://keefe.cc/stuff-alarm or in the down-loadable zip file as stuff_alarm.ino.

```
/*
 Analog Input
 http://www.arduino.cc/en/Tutorial/AnalogInput
 Created by David Cuartielles
 modified 30 Aug 2011 by Tom Igoe
 modified March 2016 by John Keefe
 This example code is in the public domain.
 */
int sensorPin = A0;    // set the input pin for the sensor
int buzzerPin = 13;    // set the pin for the buzzer
int sensorValue = 0;   // variable to store the sensor value
int movedValue = 100;  // threshold value to trigger the buzzer
void setup() {
  // declare the buzzerPin as an OUTPUT:
  pinMode(buzzerPin, OUTPUT);
  // start up the Serial Monitor
  Serial.begin(9600);
}
void loop() {
  // read the value from the sensor:
  sensorValue = analogRead(sensorPin);

  // if the sensorValue is less than the movedValue,
  // turn the buzzer on (with HIGH power).
  // Otherwise turn it off (with LOW power).
  if (sensorValue < movedValue) {
    digitalWrite(buzzerPin, HIGH);
  } else {
    digitalWrite(buzzerPin, LOW);
  }
  // print the sensor value to the Serial Monitor for calibration
  Serial.println(sensorValue);
}
```

ELECTRIC CANDLE

This program is also available at http://keefe.cc/electric-candle or in the downloadable zip file as electric_candle.ino.

This sketch is derived from example code provided by Modern Device for use with their hardware.

```
/* Modern Device Wind Sensor Sketch for Rev C Wind Sensor
 *
 *
Hardware Setup:
Wind Sensor Signals     Arduino
GND                     GND
+V                      5V
RV                      A1    // modify the definitions below to use other pins
TMP                     A0    // modify the definitions below to use other pins

Paul Badger 2014
Original at https://github.com/moderndevice/Wind_Sensor
Licensed for use on official Modern Device hardware
Reproduced with the permission of Paul Badger

Revised by John Keefe 2016

Hardware setup:
Wind Sensor is powered from a regulated five volt source.
RV pin and TMP pin are connected to analog inputs.

*/
#define analogPinForRV    1    // change to pins the analog pins are using
#define analogPinForTMP   0
const float zeroWindAdjustment =  .2;
int TMP_Therm_ADunits;
float RV_Wind_ADunits;
float RV_Wind_Volts;
unsigned long lastMillis;
int TempCtimes100;
float zeroWind_ADunits;
float zeroWind_volts;
float WindSpeed_MPH;
int led = 13;             // candle LED
const int buttonPin = 2;  // the pushbutton pin
int buttonState = 0;      // variable for reading the pushbutton status
void setup() {

  // initialize the digital pin as an output.
  pinMode(led, OUTPUT);

  // initialize the pushbutton pin as an input:
```

```
    pinMode(buttonPin, INPUT);

  // turn LED on
  digitalWrite(led, HIGH);
}
void loop() {

  buttonState = digitalRead(buttonPin);
  if (millis() - lastMillis > 200){

    TMP_Therm_ADunits = analogRead(analogPinForTMP);
    RV_Wind_ADunits = analogRead(analogPinForRV);
    RV_Wind_Volts = (RV_Wind_ADunits *  0.0048828125);
    // these are all derived from regressions from raw data as such they depend on a
lot of experimental factors
    // such as accuracy of temp sensors, and voltage at the actual wind sensor,
(wire losses) which were unaccounted for.
    TempCtimes100 = (0.005 *((float)TMP_Therm_ADunits * (float)TMP_Therm_ADunits)) -
(16.862 * (float)TMP_Therm_ADunits) + 9075.4;
    zeroWind_ADunits = -0.0006*((float)TMP_Therm_ADunits * (float)TMP_Therm_ADunits)
+ 1.0727 * (float)TMP_Therm_ADunits + 47.172;  //  13.0C   553   482.39
    zeroWind_volts = (zeroWind_ADunits * 0.0048828125) - zeroWindAdjustment;

  WindSpeed_MPH =  pow(((RV_Wind_Volts - zeroWind_volts) /.2300) , 2.7265);

    if (WindSpeed_MPH > 6) {
      douseCandle();
    }

    if (buttonState == HIGH) {
      lightCandle();
    }

    lastMillis = millis();
  }
}
void douseCandle() {
  // turn LED off
  digitalWrite(led, LOW);

}
void lightCandle() {
  digitalWrite(led, HIGH);

}
```

INVISIBLE RULER

This sketch is also available at http://keefe.cc/invisible ruler or in the downloadable zip file as `invisible_ruler.ino`.

```
/* Ping))) Sensor
   http://www.arduino.cc/en/Tutorial/Ping

   created 3 Nov 2008
   by David A. Mellis
   modified 30 Aug 2011
   by Tom Igoe

   This example code is in the public domain.
 */
const int pingPin = 7;
void setup() {
  // initialize serial communication:
  Serial.begin(9600);
}
void loop()
{
  // establish variables for duration of the ping,
  // and the distance result in inches and centimeters:
  long duration, inches, cm;
  // The PING))) is triggered by a HIGH pulse of 2 or more microseconds.
  // Give a short LOW pulse beforehand to ensure a clean HIGH pulse:
  pinMode(pingPin, OUTPUT);
  digitalWrite(pingPin, LOW);
  delayMicroseconds(2);

  // the next three lines send out a 5-microsecond "chirp"
  digitalWrite(pingPin, HIGH);
  delayMicroseconds(5);
  digitalWrite(pingPin, LOW);
  // The same pin is used to read the signal from the PING))): a HIGH
  // pulse whose duration is the time (in microseconds) from the sending
  // of the ping to the reception of its echo off of an object.
  pinMode(pingPin, INPUT);
  duration = pulseIn(pingPin, HIGH);
  // convert the time into a distance
  inches = microsecondsToInches(duration);
  cm = microsecondsToCentimeters(duration);
```

```
  Serial.print(inches);
  Serial.print("in, ");
  Serial.print(cm);
  Serial.print("cm");
  Serial.println();

  delay(100);
}
long microsecondsToInches(long microseconds)
{
  // According to Parallax's datasheet for the PING))), there are
  // 73.746 microseconds per inch (i.e. sound travels at 1130 feet per
  // second). This gives the distance traveled by the ping, outbound
  // and return, so we divide by 2 to get the distance of the obstacle.
  // See: http://www.parallax.com/dl/docs/prod/acc/28015-PING-v1.3.pdf
  return microseconds / 74 / 2;
}
long microsecondsToCentimeters(long microseconds)
{
  // The speed of sound is 340 m/s or 29 microseconds per centimeter.
  // The ping travels out and back, so to find the distance of the
  // object we take half of the distance traveled.
  return microseconds / 29 / 2;
}
```

GET YOUR ARDUINO ONLINE

This example code comes with the wifi shield library used in this chapter. For details on downloading the library, see the chapter or visit http://keefe.cc/arduino-online.

Once in place, you can find the code from the Arduino software menu bar; it's at *File* ➜ *Examples* ➜ *SparkFun ESP8266 AT Library* ➜ *ESP8266_Sheild_Demo*.

This sketch is also available at http://keefe.cc/arduino-online or in the down-loadable zip file as arduino_online.ino.

```
/*************************************************************
ESP8266_Shield_Demo.h
SparkFun ESP8266 AT library - Demo
Jim Lindblom @ SparkFun Electronics
```

```
Original Creation Date: July 16, 2015
https://github.com/sparkfun/SparkFun_ESP8266_AT_Arduino_Library
This example demonstrates the basics of the SparkFun ESP8266
AT library. It'll show you how to connect to a WiFi network,
get an IP address, connect over TCP to a server (as a client),
and set up a TCP server of our own.
Development environment specifics:
  IDE: Arduino 1.6.5
  Hardware Platform: Arduino Uno
  ESP8266 WiFi Shield Version: 1.0
This code is released under the MIT license.
Distributed as-is; no warranty is given.
*************************************************************/
///////////////////////
// Library Includes //
///////////////////////
// SoftwareSerial is required (even you don't intend on
// using it).
#include <SoftwareSerial.h>
#include <SparkFunESP8266WiFi.h>
//////////////////////////////
// WiFi Network Definitions //
//////////////////////////////
// Replace these two character strings with the name and
// password of your WiFi network.
const char mySSID[] = "yourSSIDhere";
const char myPSK[] = "yourPWDhere";
//////////////////////////////
// ESP8266Server definition //
//////////////////////////////
// server object used towards the end of the demo.
// (This is only global because it's called in both setup()
// and loop()).
ESP8266Server server = ESP8266Server(80);
//////////////////
// HTTP Strings //
//////////////////
const char destServer[] = "example.com";
const String htmlHeader = "HTTP/1.1 200 OK\r\n"
                          "Content-Type: text/html\r\n"
                          "Connection: close\r\n\r\n"
                          "<!DOCTYPE HTML>\r\n"
                          "<html>\r\n";
const String httpRequest = "GET / HTTP/1.1\n"
```

```
                           "Host: example.com\n"
                           "Connection: close\n\n";
// All functions called from setup() are defined below the
// loop() function. They are modularized to make it easier to
// copy/paste into sketches of your own.
void setup()
{
  // Serial Monitor is used to control the demo and view
  // debug information.
  Serial.begin(9600);
  serialTrigger(F("Press any key to begin."));
  // initializeESP8266() verifies communication with the WiFi
  // shield, and sets it up.
  initializeESP8266();
  // connectESP8266() connects to the defined WiFi network.
  connectESP8266();
  // displayConnectInfo prints the Shield's local IP
  // and the network it's connected to.
  displayConnectInfo();
  serialTrigger(F("Press any key to connect client."));
  clientDemo();

  serialTrigger(F("Press any key to test server."));
  serverSetup();
}
void loop()
{
  serverDemo();
}
void initializeESP8266()
{
  // esp8266.begin() verifies that the ESP8266 is operational
  // and sets it up for the rest of the sketch.
  // It returns either true or false -- indicating whether
  // communication was successful or not.
  // true
  int test = esp8266.begin();
  if (test != true)
  {
    Serial.println(F("Error talking to ESP8266."));
    errorLoop(test);
  }
  Serial.println(F("ESP8266 Shield Present"));
}
```

```
void connectESP8266()
{
  // The ESP8266 can be set to one of three modes:
  //  1 - ESP8266_MODE_STA - Station only
  //  2 - ESP8266_MODE_AP - Access point only
  //  3 - ESP8266_MODE_STAAP - Station/AP combo
  // Use esp8266.getMode() to check which mode it's in:
  int retVal = esp8266.getMode();
  if (retVal != ESP8266_MODE_STA)
  { // If it's not in station mode.
    // Use esp8266.setMode([mode]) to set it to a specified
    // mode.
    retVal = esp8266.setMode(ESP8266_MODE_STA);
    if (retVal < 0)
    {
      Serial.println(F("Error setting mode."));
      errorLoop(retVal);
    }
  }
  Serial.println(F("Mode set to station"));
  // esp8266.status() indicates the ESP8266's WiFi connect
  // status.
  // A return value of 1 indicates the device is already
  // connected. 0 indicates disconnected. (Negative values
  // equate to communication errors.)
  retVal = esp8266.status();
  if (retVal <= 0)
  {
    Serial.print(F("Connecting to "));
    Serial.println(mySSID);
    // esp8266.connect([ssid], [psk]) connects the ESP8266
    // to a network.
    // On success the connect function returns a value >0
    // On fail, the function will either return:
    //  -1: TIMEOUT - The library has a set 30s timeout
    //  -3: FAIL - Couldn't connect to network.
    retVal = esp8266.connect(mySSID, myPSK);
    if (retVal < 0)
    {
      Serial.println(F("Error connecting"));
      errorLoop(retVal);
    }
  }
}
```

```
void displayConnectInfo()
{
  char connectedSSID[24];
  memset(connectedSSID, 0, 24);
  // esp8266.getAP() can be used to check which AP the
  // ESP8266 is connected to. It returns an error code.
  // The connected AP is returned by reference as a parameter.
  int retVal = esp8266.getAP(connectedSSID);
  if (retVal > 0)
  {
    Serial.print(F("Connected to: "));
    Serial.println(connectedSSID);
  }
  // esp8266.localIP returns an IPAddress variable with the
  // ESP8266's current local IP address.
  IPAddress myIP = esp8266.localIP();
  Serial.print(F("My IP: ")); Serial.println(myIP);
}
void clientDemo()
{
  // To use the ESP8266 as a TCP client, use the
  // ESP8266Client class. First, create an object:
  ESP8266Client client;
  // ESP8266Client connect([server], [port]) is used to
  // connect to a server (const char * or IPAddress) on
  // a specified port.
  // Returns: 1 on success, 2 on already connected,
  // negative on fail (-1=TIMEOUT, -3=FAIL).
  int retVal = client.connect(destServer, 80);
  if (retVal <= 0)
  {
    Serial.println(F("Failed to connect to server."));
    return;
  }
  // print and write can be used to send data to a connected
  // client connection.
  client.print(httpRequest);
  // available() will return the number of characters
  // currently in the receive buffer.
  while (client.available())
    Serial.write(client.read()); // read() gets the FIFO char

  // connected() is a boolean return value - 1 if the
  // connection is active, 0 if it's closed.
```

```
      if (client.connected())
        client.stop(); // stop() closes a TCP connection.
    }
    void serverSetup()
    {
      // begin initializes a ESP8266Server object. It will
      // start a server on the port specified in the object's
      // constructor (in global area)
      server.begin();
      Serial.print(F("Server started! Go to "));
      Serial.println(esp8266.localIP());
      Serial.println();
    }
    void serverDemo()
    {
      // available() is an ESP8266Server function which will
      // return an ESP8266Client object for printing and reading.
      // available() has one parameter -- a timeout value. This
      // is the number of milliseconds the function waits,
      // checking for a connection.
      ESP8266Client client = server.available(500);

      if (client)
      {
        Serial.println(F("Client Connected!"));
        // an http request ends with a blank line
        boolean currentLineIsBlank = true;
        while (client.connected())
        {
          if (client.available())
          {
            char c = client.read();
            // if you've gotten to the end of the line (received a newline
            // character) and the line is blank, the http request has ended,
            // so you can send a reply
            if (c == '\n' && currentLineIsBlank)
            {
              Serial.println(F("Sending HTML page"));
              // send a standard http response header:
              client.print(htmlHeader);
              String htmlBody;
              // output the value of each analog input pin
              for (int a = 0; a < 6; a++)
              {
```

```
            htmlBody += "A";
            htmlBody += String(a);
            htmlBody += ": ";
            htmlBody += String(analogRead(a));
            htmlBody += "<br>\n";
          }
          htmlBody += "</html>\n";
          client.print(htmlBody);
          break;
        }
        if (c == '\n')
        {
          // you're starting a new line
          currentLineIsBlank = true;
        }
        else if (c != '\r')
        {
          // you've gotten a character on the current line
          currentLineIsBlank = false;
        }
      }
    }
    // give the web browser time to receive the data
    delay(1);

    // close the connection:
    client.stop();
    Serial.println(F("Client disconnected"));
  }

}
// errorLoop prints an error code, then loops forever.
void errorLoop(int error)
{
  Serial.print(F("Error: ")); Serial.println(error);
  Serial.println(F("Looping forever."));
  for (;;)
    ;
}
// serialTrigger prints a message, then waits for something
// to come in from the serial port.
void serialTrigger(String message)
{
  Serial.println();
```

```
   Serial.println(message);
   Serial.println();
   while (!Serial.available())
      ;
   while (Serial.available())
      Serial.read();
}
```

DO I NEED AN UMBRELLA TODAY?

This sketch is also available at http://keefe.cc/umbrella-today or in the downloadable zip file as umbrella_today.ino.

```
/*************************************************************
ESP8266_Shield_Demo.h
SparkFun ESP8266 AT library - Demo
Jim Lindblom @ SparkFun Electronics
Original Creation Date: July 16, 2015
https://github.com/sparkfun/SparkFun_ESP8266_AT_Arduino_Library
This example demonstrates the basics of the SparkFun ESP8266
AT library. It'll show you how to connect to a WiFi network,
get an IP address, connect over TCP to a server (as a client),
and set up a TCP server of our own.
Development environment specifics:
   IDE: Arduino 1.6.5
   Hardware Platform: Arduino Uno
   ESP8266 WiFi Shield Version: 1.0
This code is released under the MIT license.
Distributed as-is; no warranty is given.
---
Modified by John Keefe
May 2016
To check OpenWeatherMap for rain forecast in next 12-15 hrs
   If LED in Pin 13 & ground ...
   Quick "heartbeat" flashes during check,
   Steady = No rain in next 12-15 hrs
   Flashing = Bring an umbrella!

   Reset or turn off/on to check again
This code is released under the MIT license.
Distributed as-is; no warranty is given.
*************************************************************/
```

```
/////////////////////
// Library Includes //
/////////////////////
// SoftwareSerial is required (even you don't intend on
// using it).
#include <SoftwareSerial.h>
#include <SparkFunESP8266WiFi.h>
/////////////////////////
// Weather Variables //
/////////////////////////
const String apikey = "YourAPIKeyGoesHere";
const String latitude = "YourLatitudeGoesHere";
const String longitude = "YourLongitudeGoesHere";
/////////////////////////////////
// WiFi Network Definitions //
/////////////////////////////////
const char mySSID[] = "YourWifiNetworkNameGoesHere";
const char myPSK[] = "YourWifiPasswordGoesHere";
/////////////////////////////
// Forecast Variables //
/////////////////////////////
boolean forecastFound = false;
boolean itWillRain = false;
////////////////////
// HTTP Strings //
////////////////////
const char destServer[] = "api.openweathermap.org";
const String httpRequest = "GET /data/2.5/forecast?lat=" + latitude + "&lon=" +
longitude +
                          "&APPID=" + apikey + " HTTP/1.1\n"
                          "Host: api.openweathermap.org\n"
                          "Connection: close\n\n";

// All functions called from setup() are defined below the
// loop() function. They are modularized to make it easier to
// copy/paste into sketches of your own.
void setup()
{
  // Initialize digital pin 13 as an output.
  pinMode(13, OUTPUT);

  // Serial Monitor is used to control the demo and view
  // debug information.
  Serial.begin(9600);
```

```
    // initializeESP8266() verifies communication with the WiFi
    // shield, and sets it up.
    initializeESP8266();
    // connectESP8266() connects to the defined WiFi network.
    connectESP8266();
    // displayConnectInfo prints the Shield's local IP
    // and the network it's connected to.
    displayConnectInfo();
    // go check the forecast!
    delay(100);
    getWeather();

    // adding an extra blank line for clarity
    Serial.println();


    // finally, let us know if it's going to rain in the next 12-15 hours.
    // note that because itWillRain is a true/false (or boolean) variable,
    if (itWillRain == true) {
      Serial.println("Better bring an umbrella: Rain forecast in the next 12-15
hours.");
      // flash a rain warning!
      for (int i = 0; i <  20; i++){
        digitalWrite(13, HIGH);    // turn the LED on (HIGH is the voltage level)
        delay(1000);               // wait for a second
        digitalWrite(13, LOW);     // turn the LED off by making the voltage LOW
        delay(1000);               // wait for a second
      }
    } else if (itWillRain == false && forecastFound == true) {
      Serial.println("No rain in the forecast for the next 12-15 hours.");
      // give a steady all-clear
      digitalWrite(13, HIGH);    // turn the LED on (HIGH is the voltage level)
      delay(20000);              // wait for 20 seconds
      digitalWrite(13, LOW);     // turn the LED off by making the voltage LOW
    } else {
      // forecastFound is still false
      Serial.println("Hmmmm. Having trouble reading the forecast :-(");
    }

}
void loop()
{
  // nothing here
}
void initializeESP8266()
```

```
{
  // esp8266.begin() verifies that the ESP8266 is operational
  // and sets it up for the rest of the sketch.
  // It returns either true or false -- indicating whether
  // communication was successful or not.
  // true
  int test = esp8266.begin();
  if (test != true)
  {
    Serial.println(F("Error talking to ESP8266."));
    errorLoop(test);
  }
  Serial.println(F("ESP8266 Shield Present"));
}
void connectESP8266()
{
  // The ESP8266 can be set to one of three modes:
  //   1 - ESP8266_MODE_STA - Station only
  //   2 - ESP8266_MODE_AP - Access point only
  //   3 - ESP8266_MODE_STAAP - Station/AP combo
  // Use esp8266.getMode() to check which mode it's in:
  int retVal = esp8266.getMode();
  if (retVal != ESP8266_MODE_STA)
  { // If it's not in station mode.
    // Use esp8266.setMode([mode]) to set it to a specified
    // mode.
    retVal = esp8266.setMode(ESP8266_MODE_STA);
    if (retVal < 0)
    {
      Serial.println(F("Error setting mode."));
      errorLoop(retVal);
    }
  }
  Serial.println(F("Mode set to station"));
  // esp8266.status() indicates the ESP8266's WiFi connect
  // status.
  // A return value of 1 indicates the device is already
  // connected. 0 indicates disconnected. (Negative values
  // equate to communication errors.)
  retVal = esp8266.status();
  if (retVal <= 0)
  {
    Serial.print(F("Connecting to "));
    Serial.println(mySSID);
```

```
    // esp8266.connect([ssid], [psk]) connects the ESP8266
    // to a network.
    // On success the connect function returns a value >0
    // On fail, the function will either return:
    //   -1: TIMEOUT - The library has a set 30s timeout
    //   -3: FAIL - Couldn't connect to network.
    retVal = esp8266.connect(mySSID, myPSK);
    if (retVal < 0)
    {
      Serial.println(F("Error connecting"));
      errorLoop(retVal);
    }
  }
}
void displayConnectInfo()
{
  char connectedSSID[24];
  memset(connectedSSID, 0, 24);
  // esp8266.getAP() can be used to check which AP the
  // ESP8266 is connected to. It returns an error code.
  // The connected AP is returned by reference as a parameter.
  int retVal = esp8266.getAP(connectedSSID);
  if (retVal > 0)
  {
    Serial.print(F("Connected to: "));
    Serial.println(connectedSSID);
  }
  // esp8266.localIP returns an IPAddress variable with the
  // ESP8266's current local IP address.
  IPAddress myIP = esp8266.localIP();
  Serial.print(F("My IP: ")); Serial.println(myIP);
}
void getWeather()
{
  // To use the ESP8266 as a TCP client, use the
  // ESP8266Client class. First, create an object:
  ESP8266Client client;
  // ESP8266Client connect([server], [port]) is used to
  // connect to a server (const char * or IPAddress) on
  // a specified port.
  // Returns: 1 on success, 2 on already connected,
  // negative on fail (-1=TIMEOUT, -3=FAIL).
  int retVal = client.connect(destServer, 80);
  if (retVal <= 0)
```

```
      {
        Serial.println(F("Failed to connect to server."));
        return;
      }
      // These are variables we're going to use below. Read on!
      int numberOfFinds = 0;
      // print and write can be used to send data to a connected
      // client connection. Here, it sends the "request," which
      // is the information we're asking for.
      client.print(httpRequest);
      // Here's where we read the data from the website.
      // available() will return the number of characters
      // currently in the receive buffer.
      while (client.available()) {
        // Is it raining in the next 12-15 hours? Let's find out.
        // Openweathermap.org sends the 5-day forecast in a stream of data
        // formatted as seen here: http://openweathermap.org/forecast5#JSON.
        // Here's a full sample: http://keefe.cc/sensing-internet/sample-data.json

        // In that stream, there are 40 forecasts, one for every 3 hours
        // over the next 5 days. So for the next 12-15 hours, we want to check
        // the first five forecasts.

        // Each forecast has an "id" that indicates
        // the type of weather forecast. See chart here:
        // http://openweathermap.org/weather-conditions
        // From the chart, we see that if the "id" starts with a 3 or a 5,
        // it's going to rain (or at least drizzle). The data looks like this:
        //     ... "weather":[{"id":500,"main": ...
        // Using client.find("<something to find>"), we'll look for
        // the first five appearances of  "id": ... with the quotes and the colon
        // ...  and then read the very next character to see if it's a 3 or a 5.
        // There's actually one other "id": at the beginning of the stream that is
        // used to identify the city, so we'll ignore the first case, using cases 2
through 6.
        // Also, when using the client.find() command, we have to wrap the
        // search part in double quotes, like client.find("my search"). But the
        // thing we're searching for actually *includes* double quotes: "id":
        // ... and client.find(""id"":) will totally confuse the system.
        // So I use the "hex" value for a double quote, 22, and
        // represent each of them in Arduino-speak as \x22.
        // so "id": becomes \x22id\x22:
        // look for "id": in the stream of data, written as \x22id\x22:
        if ( client.find( "\x22id\x22:" ) ) {
```

```
    // found one! Note that and add to the number
    forecastFound = true;
    numberOfFinds += 1;
    Serial.print("id number ");
    Serial.print(numberOfFinds);
    Serial.print(" ... ");
    // flicker the LED on the first find and every 10 finds
    // so we know it's working
    if (numberOfFinds == 1 || numberOfFinds % 10 == 0) {
      digitalWrite(13, HIGH);   // turn the LED on (HIGH is the voltage level)
      delay(2);                 // wait for a 10th of a second
      digitalWrite(13, LOW);    // turn the LED off by making the voltage LOW
    }

    // remember, we only want the 2nd through 6th "id"
    // so looking for the ones greater than 1 and less
    // than 7. The && means "and."
    if ( numberOfFinds > 1 && numberOfFinds < 7) {
      // read the next character
      char nextCharacter = client.read();

      // and convert it into a number
      int nextNumber = (int)nextCharacter - 48;
      Serial.print("next number is ");
      Serial.print(nextNumber);
      Serial.print(" ... ");
      // is it a 3 or a 5?
      // (the double bars || mean "or")
      if (nextNumber == 3 || nextNumber == 5) {
        // found a 3 or a 5
        itWillRain = true;
        Serial.println("Rain forecast!");
      } else {
        // not a 3 or a 5
        Serial.println("No rain!");
      }

    } else {
      // number of finds wasn't greater than 1 and less than 7
      Serial.println("skipping!");
    }

  }
```

```
  }
  // This code runs when the connection to the website is done.
  // connected() is a boolean return value - 1 if the
  // connection is active, 0 if it's closed.
  if (client.connected()) {
    client.stop(); // stop() closes a TCP connection.
  }
}
// errorLoop prints an error code, then loops forever.
void errorLoop(int error)
{
  Serial.print(F("Error: ")); Serial.println(error);
  Serial.println(F("Looping forever."));
  for (;;)
    ;
}
```

SEND EMAIL WITH A BUTTON

This sketch is also available at http://keefe.cc/email-button or in the downloadable zip file as email_button.ino.

```
/*************************************************************
Built on example code ESP8266_Shield_Demo.h
SparkFun ESP8266 AT library - Demo
Jim Lindblom @ SparkFun Electronics
Original Creation Date: July 16, 2015
https://github.com/sparkfun/SparkFun_ESP8266_AT_Arduino_Library
This example demonstrates the basics of the SparkFun ESP8266
AT library. It'll show you how to connect to a WiFi network,
get an IP address, connect over TCP to a server (as a client),
and set up a TCP server of our own.
Development environment specifics:
  IDE: Arduino 1.6.5
  Hardware Platform: Arduino Uno
  ESP8266 WiFi Shield Version: 1.0
This code is released under the MIT license.
Distributed as-is; no warranty is given.
*******
Also includes "Button" example code:
 Turns on and off a light emitting diode(LED) connected to digital
 pin 13, when pressing a pushbutton attached to pin 2.
```

```
   The circuit:
   * LED attached from pin 13 to ground
   * pushbutton attached to pin 2 from +5V
   * 10K resistor attached to pin 2 from ground
   created 2005
   by DojoDave <http://www.0j0.org>
   modified 30 Aug 2011
   by Tom Igoe
   This example code is in the public domain.
   http://www.arduino.cc/en/Tutorial/Button
****
Updated for IFTTT
by John Keefe
http://johnkeefe.net
June 2016
*****************************************************************/
///////////////////////
// Library Includes //
///////////////////////
// SoftwareSerial is required (even you don't intend on
// using it).
#include <SoftwareSerial.h>
#include <SparkFunESP8266WiFi.h>
/////////////////////////////////
// WiFi Network Definitions //
/////////////////////////////////
// const char mySSID[] = "YourWifiNetworkNameGoesHere";
// const char myPSK[] = "YourWifiPasswordGoesHere";
/////////////////////////
// IFTTT Variable //
/////////////////////////
//const String maker_key = "YourMakerKeyGoesHere";
/////////////////////////////////////
// Other Constants & Variables //
/////////////////////////////////////
const int buttonPin = 2;   // the number of the pushbutton pin
const int ledPin =  13;    // the number of the LED pin
int buttonState = 0;       // variable for reading the pushbutton status
/////////////////////
// HTTP Strings //
/////////////////////
const String iftttServer = "maker.ifttt.com";
String httpHeader = "GET /trigger/button_pressed/with/key/" + maker_key + "
HTTP/1.1\r\n" +
```

```
                        "Host: " + iftttServer + "\r\n" +
                        "User-Agent: Arduino/1.0\r\n" +
                        "Connection: close\r\n";


// All functions called from setup() are defined below the
// loop() function. They are modularized to make it easier to
// copy/paste into sketches of your own.
void setup()
{
  // initialize the LED pin as an output:
  pinMode(ledPin, OUTPUT);

  // initialize the pushbutton pin as an input:
  pinMode(buttonPin, INPUT);

  // Serial Monitor is used to control the demo and view
  // debug information.
  Serial.begin(9600);
  // initializeESP8266() verifies communication with the WiFi
  // shield, and sets it up.
  initializeESP8266();
  // connectESP8266() connects to the defined WiFi network.
  connectESP8266();
  // displayConnectInfo prints the Shield's local IP
  // and the network it's connected to.
  displayConnectInfo();
  // adding an extra blank line for clarity
  Serial.println();
  // turn LED on
  digitalWrite(ledPin, HIGH);
  // This allows the wifi connection to get stable after starting
  // or hitting reset
  waitAMinute();

}
void loop() {

  // douse the LED
  digitalWrite(ledPin, LOW);

  // read the state of the pushbutton value:
  buttonState = digitalRead(buttonPin);
  // check if the pushbutton is pressed.
  // if it is, the buttonState is HIGH:
```

```
  if (buttonState == HIGH) {

    // turn LED on:
    digitalWrite(ledPin, HIGH);

    // send trigger to IFTTT
    sendTrigger();
    // do nothing for a full minute
    // this ensures you don't accidentally send yourself 100s of emails!
    waitAMinute();

  }
}
void waitAMinute()
{
  Serial.println("Waiting a full minute to run.");
  delay(60000);
  Serial.println("Ready ... ");
}
void initializeESP8266()
{
  // esp8266.begin() verifies that the ESP8266 is operational
  // and sets it up for the rest of the sketch.
  // It returns either true or false -- indicating whether
  // communication was successful or not.
  // true
  int test = esp8266.begin();
  if (test != true)
  {
    Serial.println(F("Error talking to ESP8266."));
    errorLoop(test);
  }
  Serial.println(F("ESP8266 Shield Present"));
}
void connectESP8266()
{
  // The ESP8266 can be set to one of three modes:
  //   1 - ESP8266_MODE_STA - Station only
  //   2 - ESP8266_MODE_AP - Access point only
  //   3 - ESP8266_MODE_STAAP - Station/AP combo
  // Use esp8266.getMode() to check which mode it's in:
  int retVal = esp8266.getMode();
  if (retVal != ESP8266_MODE_STA)
  { // If it's not in station mode.
```

```
    // Use esp8266.setMode([mode]) to set it to a specified
    // mode.
    retVal = esp8266.setMode(ESP8266_MODE_STA);
    if (retVal < 0)
    {
      Serial.println(F("Error setting mode."));
      errorLoop(retVal);
    }
  }
  Serial.println(F("Mode set to station"));
  // esp8266.status() indicates the ESP8266's WiFi connect
  // status.
  // A return value of 1 indicates the device is already
  // connected. 0 indicates disconnected. (Negative values
  // equate to communication errors.)
  retVal = esp8266.status();
  if (retVal <= 0)
  {
    Serial.print(F("Connecting to "));
    Serial.println(mySSID);
    // esp8266.connect([ssid], [psk]) connects the ESP8266
    // to a network.
    // On success the connect function returns a value >0
    // On fail, the function will either return:
    //   -1: TIMEOUT - The library has a set 30s timeout
    //   -3: FAIL - Couldn't connect to network.
    retVal = esp8266.connect(mySSID, myPSK);
    if (retVal < 0)
    {
      Serial.println(F("Error connecting"));
      errorLoop(retVal);
    }
  }
}
void displayConnectInfo()
{
  char connectedSSID[24];
  memset(connectedSSID, 0, 24);
  // esp8266.getAP() can be used to check which AP the
  // ESP8266 is connected to. It returns an error code.
  // The connected AP is returned by reference as a parameter.
  int retVal = esp8266.getAP(connectedSSID);
  if (retVal > 0)
  {
```

```
    Serial.print(F("Connected to: "));
    Serial.println(connectedSSID);
  }
  // esp8266.localIP returns an IPAddress variable with the
  // ESP8266's current local IP address.
  IPAddress myIP = esp8266.localIP();
  Serial.print(F("My IP: ")); Serial.println(myIP);
}
void sendTrigger()
{
  // Create a client, and initiate a connection
  ESP8266Client client;

  if (client.connect(iftttServer, 80) <= 0)
  {
    Serial.println(F("Failed to connect to server."));
    return;
  }
  Serial.println(F("Connected."));

  Serial.println(F("Posting to IFTTT!"));
  client.print(httpHeader);
  client.println();
  Serial.println(httpHeader);
  // available() will return the number of characters
  // currently in the receive buffer.
  while (client.available())
    Serial.write(client.read()); // read() gets the FIFO char

  // connected() is a boolean return value - 1 if the
  // connection is active, 0 if it's closed.
  if (client.connected())
    client.stop(); // stop() closes a TCP connection.
}
// errorLoop prints an error code, then loops forever.
void errorLoop(int error)
{
  Serial.print(F("Error: ")); Serial.println(error);
  Serial.println(F("Looping forever."));
  for (;;)
    ;
}
```

ONLINE TEMPERATURE TRACKER

This sketch is also available at http://keefe.cc/temp-tracker or in the downloadable zip file as `temp_tracker.ino`.

```
/****************************************************************
This is a modified version of ESP8266_Phant.ino
SparkFun ESP8266 AT library - Phant Posting Example
Jim Lindblom @ SparkFun Electronics
Original Creation Date: July 16, 2015
https://github.com/sparkfun/SparkFun_ESP8266_AT_Arduino_Library
This example demonstrates how to use the TCP client
functionality of the SparkFun ESP8266 WiFi library to post
sensor readings to a Phant stream on
https://data.sparkfun.com
This sketch is set up to post to a publicly available stream
https://data.sparkfun.com/streams/DJjNowwjgxFR9ogvr45Q
Please don't abuse it! But feel free to post a few times to
verify the sketch works. If it fails, check the HTTP response
to make sure the post rate hasn't been exceeded.
Development environment specifics:
  IDE: Arduino 1.6.5
  Hardware Platform: Arduino Uno
  ESP8266 WiFi Shield Version: 1.0
This code is beerware; if you see me (or any other SparkFun
employee) at the local, and you've found our code helpful,
please buy us a round!
Distributed as-is; no warranty is given.
---
Modified for a temperature sensor
by John Keefe
June 2016
****************************************************************/
// The SparkFunESP8266WiFi library uses SoftwareSerial
// to communicate with the ESP8266 module. Include that
// library first:
#include <SoftwareSerial.h>
// Include the ESP8266 AT library:
#include <SparkFunESP8266WiFi.h>
#define SENSORPIN 0
/////////////////////////////
// WiFi Network Definitions //
/////////////////////////////
```

```
// const char mySSID[] = "YourWifiNetworkNameGoesHere";
// const char myPSK[] = "YourWifiPasswordGoesHere";
/////////////////////////
// SparkFun Data Keys //
/////////////////////////
// const String publicKey = "YourPublicKeyHere";
// const String privateKey = "YourPrivateKeyHere";
// Phant destination server:
const String phantServer = "data.sparkfun.com";
String httpHeader = "POST /input/" + publicKey + ".txt HTTP/1.1\n" +
                    "Host: " + phantServer + "\n" +
                    "Phant-Private-Key: " + privateKey + "\n" +
                    "Connection: close\n" +
                    "Content-Type: application/x-www-form-urlencoded\n";
/////////////////////
// Sensor Values //
/////////////////////
int waitMinutes = 2;
float sensor_value;
float voltage;
float temp;
void setup()
{
  int status;
  Serial.begin(9600);

  // To turn the MG2639 shield on, and verify communication
  // always begin a sketch by calling cell.begin().
  status = esp8266.begin();
  if (status <= 0)
  {
    Serial.println(F("Unable to communicate with shield. Looping forever."));
    while(1) ;
  }

  esp8266.setMode(ESP8266_MODE_STA); // Set WiFi mode to station
  if (esp8266.status() <= 0) // If we're not already connected
  {
    if (esp8266.connect(mySSID, myPSK) < 0)
    {
      Serial.println(F("Error connecting"));
      while (1) ;
    }
  }
```

```
    // Get our assigned IP address and print it:
    Serial.print(F("My IP address is: "));
    Serial.println(esp8266.localIP());
    Serial.println(F("Waiting one minute to get started ..."));
    delay(60000);

}
void loop()
{
  Serial.println(F("Latest reading ..."));
  sensor_value = analogRead(SENSORPIN);
  Serial.print(F("Sensor Value: "));
  Serial.println(sensor_value);

  voltage = sensor_value * 5000 / 1024;
  Serial.print(F("Voltage (in mV): "));
  Serial.println(voltage);

  temp = (voltage-500) / 10;
  Serial.print(F("Degrees Celsius: "));
  Serial.println(temp);
  postToPhant(temp);
  Serial.println(F("Done! See the current data posted online at:"));
  Serial.print  (F("https://data.sparkfun.com/streams/"));
  Serial.println(publicKey);
  Serial.println(F("Waiting for next reading ..."));
  Serial.println();
  delay(waitMinutes * 60000);

}
void postToPhant(float temperature)
{
  // Create a client, and initiate a connection
  ESP8266Client client;

  if (client.connect(phantServer, 80) <= 0)
  {
    Serial.println(F("Failed to connect to server."));
    return;
  }
  Serial.println(F("Connected."));

  // Set up our Phant post parameters:
```

```
String params;
params += "temp=" + String(temperature);

Serial.println(F("Posting to SparkFun!"));
client.print(httpHeader);
client.print("Content-Length: ");
client.println(params.length());
client.println();
client.print(params);
// available() will return the number of characters
// currently in the receive buffer.
while (client.available())
  Serial.write(client.read()); // read() gets the FIFO char

// connected() is a boolean return value - 1 if the
// connection is active, 0 if it's closed.
if (client.connected())
  client.stop(); // stop() closes a TCP connection.
}
```

Index

Number

10K-ohm (10KΩ) resistor, stripes on, 30

Symbols

/* and */ comments, using, 23, 35
+ (addition) symbol, using, 88
// (comments), using, 23
{} (curly braces) in Hello Blinky World, 22
/ (division) symbol, using, 88, 97
= (equals sign), using, 78
(minus) symbol, using, 87
* (multiplication) symbol, using, 88
Ω (ohms) symbol, 60
" (quote marks), including strings in, 111

A

Adafruit
    Arduino Lessons, 157
    soldering tutorial, 80
addition, symbol for, 88
air movement, sensing, 79–80
alarm, powering, 76
API (application programming
    interface), 115–116, 152–153. See also
    OpenWeatherMap key
Apple computers, downloading software
    for, 4–7
Arduino boards, verifying
    connection, 12–13
Arduino libraries
    CapacitiveSensor, 66–67
    for A Gentle Touch, 61–62
    for wifi board, 105–106

Arduino Online
    code, 111, 176–183
    connecting with wifi board,
        99–100
    fixes, 109–110
    function of, 110–111
    ingredients, 99
    loading up code, 105–107
    making it go, 108
    strings, 111
    test server, 112–113
    wiring up parts, 104–105
The Arduino Playground, 157
Arduino setup, fixes, 13
Arduino software downloads, 18
    for Linux users, 10–11
    for Macs, 4
    for Windows, 7
Arduino Uno
    displaying as option, 12–13
    Revision 3, 1–2
arduino-#.*.#-macosx.zip file,
    downloading and unzipping, 4–5
Arduino.app icon, displaying on
    desktop, 5
Arduinos
    forums, 24
    getting connected, 3
    for Linux users, 10–11
    for Mac OS X, 4–7
    as open source hardware, 1
    troubleshooting installation, 13
    untethering, 77
    for Windows users, 7–10

LOOK FOR MORE BOOKS FROM **Make:**

Make: Electronics, Second Edition
Learning Through Discovery
Charles Platt
US $34.99   ISBN-13: 978-1680450262

Make: Tech DIY
Easy Electronics Projects for Parents and Kids
Ji Sun Lee and Jaymes Dec
US $19.99   ISBN: 978-1680451771

Make: Action
Movement, Light, and Sound with Arduino and Raspberry Pi
Simon Monk
US $34.99   ISBN: 978-1457187797

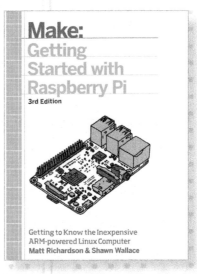

Make: Getting Started with Raspberry Pi, 3rd Edition
Getting to Know the Inexpensive ARM-Powered Linux Computer
Matt Richardson and Shawn Wallace
US $19.99   ISBN: 978-1680452464

CPSIA information can be obtained
at www.ICGtesting.com
Printed in the USA
BVOW11s1129030916

460825BV00002B/2/P